Generators in development projects

How to choose, size, install and use diesel generators economically.

Second Edition.
April 2014.

Santiago Arnalich

arnalich
water and habitat

Generators in development projects

How to choose, size, install and use diesel generators economically.

Second Edition.
Aril 2014.

ISBN: 978-1-49755-210-4

© Arnalich. Water and habitat

All rights reserved. You may photocopy this manual for your personal use if your financial situation does not allow you to buy it. Otherwise, please consider supporting these initiatives by purchasing a copy.

If you would like to use part of the contents of this book,
please contact us at publicaciones@arnalich.com.

Errata may be found at: www.arnalich.com/dwnl/xgsetcoen.doc

arnalich
water and habitat

DISCLAIMER: The information in this book has been obtained from credible and internationally respected sources. However, neither Arnalich w&h nor the authors can guarantee the accuracy of the information published here and will not be liable for any errors, omissions or damage caused by the use of this information. It is understood that information is being published without a particular purpose and is not, in any case, trying to provide professional engineering services. If such services are required, the assistance of an appropriate professional must be sought.

*To all those who suffer from energy poverty.
We hope that the following words will be translated
into livelihood improvements for some of them.*

Our gratitude to all those who, directly or indirectly, contributed to the making of this book, thanks go to:

- Ivan C. Amaya from Pramac, for his initial interest in the book that resulted in the sponsorship of this translation.
- Mary brown and Nick Gills for their proofreading.
- Special thanks to Simone Charron and Kiera Schoenemann for their effort and many suggestions on how to improve this book, and to Maxim Fortin for his help.

Index

About this book 7
The importance of generators in development projects 8
Quick anatomy of a generator 9
Avoiding confusion around the many definitions of power 11
Some basics of electricity 13

1. Economic considerations   19

Ballpark figures 20
The effect of load 22
Estimating consumption 25
Trends in the price of diesel 26
Is it worth investing in low consumption appliances? 28
Can the operating costs be covered? 31
Prioritizing the loads 33
Small, constant and important consumptions 35
Heat recovery units 38

2. Sizing   39

Types of loads 40
Estimating the power requirement 42
Derating according to the operating conditions 46
Several generators in paralell install or a single generator 47
Voltage and frequency variations with load 48
Sizing for constant consumption 49
Sizing for consumption with peak 50
Sizing of backup for power cuts 53
Sizing of a supply generator 55

3. Selection and purchase   59

Understanding purchase options and specifications 60

Ordering the generator and accessories 67
Ordering spare parts 69

4. Installation   71

Selecting the site 72
Transport of generators 73
The generator house 74
Foundation 78
Flow of air and gases 80
Fuel tanks 83
Noise reduction 85
Electrical protection 90
Transfer switches 93
Balancing the loads 96
Improving existing installations 98

5. Operation and maintenance   99

Record book 100
Measuring diesel consumption 101
Evaluating the quality of diesel 104
Overhaul 105
Long term storage 108
Operation in adverse conditions 110

Annexes   115

A. Calculation of AC power 116
B. Useful life of a generator 117
C. Grundfos submersible pump start-up peak 118
D. Approximate weights and measures of generators 119
E. Approximate dimensions of installation elements 120
F. Typical tasks involved in an overhaul 121

Bibliography 123
Pramac branches worldwide 125

About this book

This book is intended to provide you with the necessary tools and information to help you determine which generator you need, buy it, oversee its installation, organize the maintenance and get it working in the most economical way possible in a couple of evenings reading.

This book is intended to be:

99% fat free. No meticulous explanations or endless demonstrations. It only includes what you need.

Simple. A frequent cause of failure is that the complexity and excessive strictness end up being intimidating and result in things being left undone. Even at the risk of being overly simplified, the explanations try not to assume that anything is obvious.

Chronological. The book follows a roughly logical order and walks you through each step in the process. Even so, it is important to read the whole book before taking any action so as to avoid omitting important steps..

Practical. An abundance of practical examples allows you to understand the concepts and how they might relate to your own project.

We focus the on generators most commonly used in development projects: diesel generators between 5 and 200 kW. Although portable gasoline generators are very common, they do not involve the same planning difficulties.

This manual is written from the **point of view of the manager** to help him or her make informed decisions; it is not about turning you into a mechanic or an electrical fitter.

Finally, REMEMBER:

Working with electricity can be extremely dangerous and even deadly. Seek the help of local electricians and mechanics.

The importance of generators in development projects

Generators are one of the key elements in the success of a project due to the expenses they incur, the rising oil price increases and abrupt price changes caused by local conflicts or weather conditions that hinder its transport. A generator that is poorly sized or poorly planned can quickly ruin all other project efforts, and can result in a project being unsustainable for the communities it should serve.

It is relatively common to come across an unused generator which has been neatly stored away. This leads us to believe that while generators are seen as a valuable investment, they can also become unaffordable or difficult to maintain.

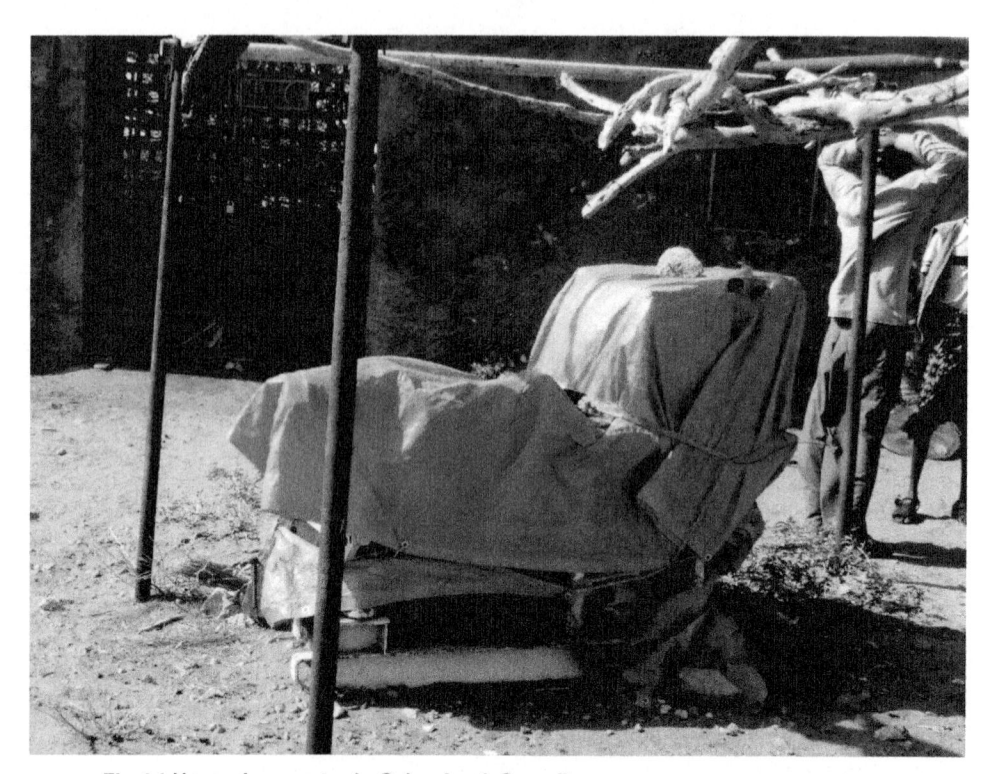

Fig. I.1 Unused generator in Galgaduud, Somalia.

When a generator is broken or neglected, you should first conduct a thorough analysis of the causes of the problem instead of rushing off to repair or replace it. Often, the problem is not the generator itself but the proposal of the generator as the solution or the approach for paying the operating expenses.

Quick anatomy of a generator

Figure I.2 shows the essential components of a generator:

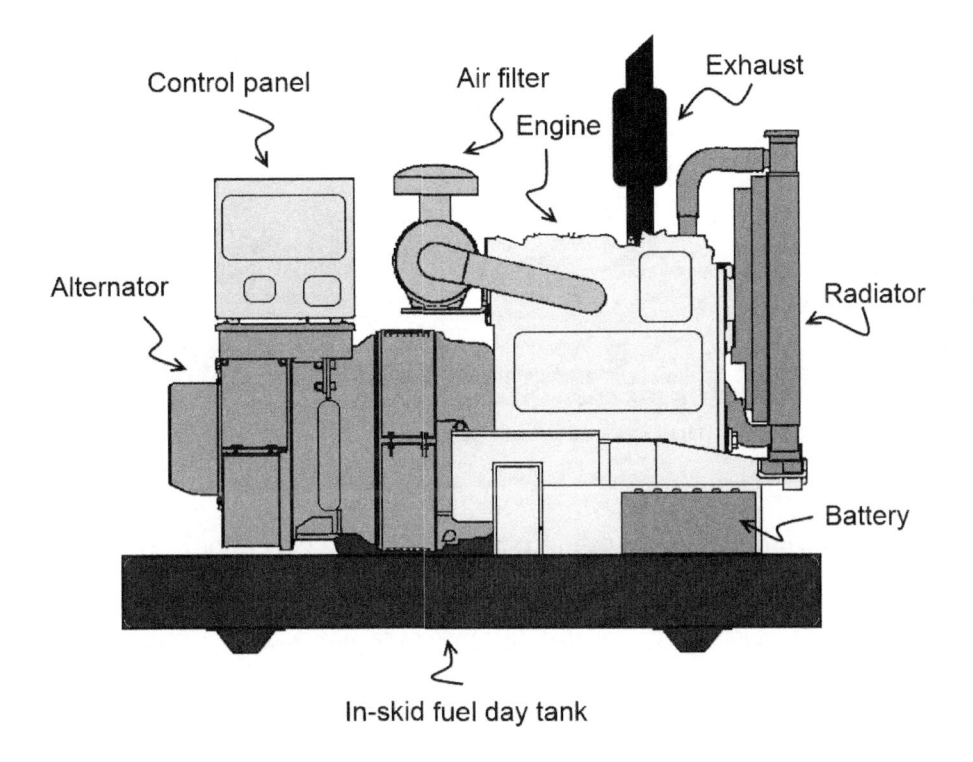

Fig. I.2 Main parts of a diesel generator.

A generator can be divided into the following components:

- **Engine.** Burns the fuel and transforms its energy into motion.
- **Alternator.** Takes the engine movement and transforms it into electricity.
- **Battery and charger.** Serves to start the engine up and feeds the control circuits during operation. Make sure the battery is always there to avoid damage!
- **Exhaust.** Evacuates the combustion gases and reduces engine noise.
- **Radiator.** Cools the engine to prevent overheating.
- **Control panel.** Where the controls and the operating indicators are found. The controls and the operating indicators and their appearance vary between panels, depending whether they are digital or analog. The basic components of the control pannel are:

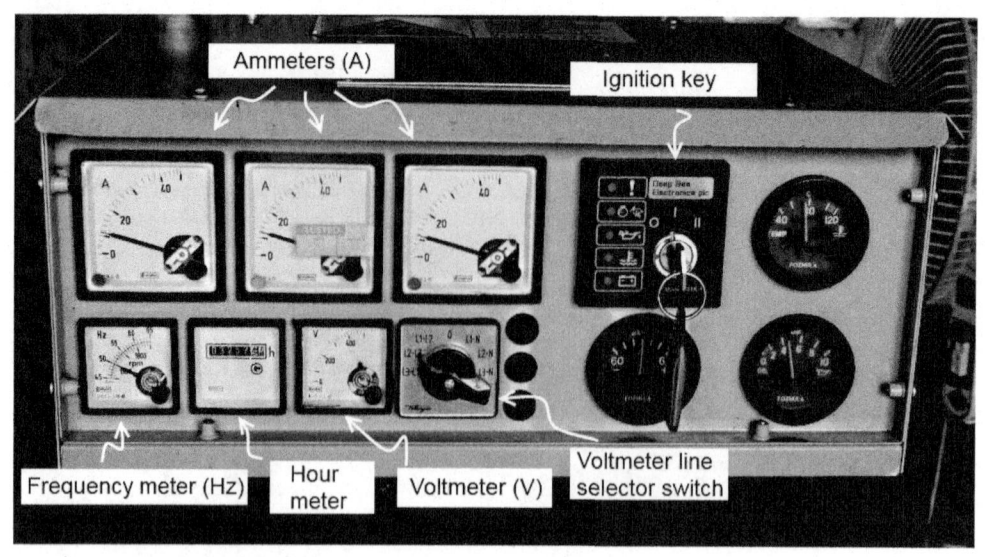

Fig. I.3 Analog control panel of a three-phase generator.

- The **ignition key** is used to start and stop the generator.
- The **ammeters** are the indicators marked with an A. They measure the amount of current circulating and there are usually 3, one for each phase.
- The **voltmeter** is recognized by the V on the display and is used to measure the voltage. The selector switch is used to determine between which lines the voltage is measured.
- The **frequency meter** is marked with Hz and measures the frequency.
- The **hour meter** is equivalent to a car's milometer. The generator maintenance is organized according to the hours in operation and serves to give an idea of the internal wear.

Avoiding confusion around the many definitions of power

Nominal power prime power, standby power, continous power ....

Generators can be specified according to different power ratings, often lending to major confusion. This means that 100 kW, 90 kW or 70 kW generators can be the same, depending on which type of power rating is involved. Also, different manufacturers will have varying interpretations or even introduce other categories. Finally, remember that power may be expressed in two different units of measurement: kW or kVA (1 kVA ≈ 0,8 kW). **Make sure that you are aware at all times of which power and which units are being utilized!**

Nominal Power (NP) or nameplate power is the maximum power which the machine provides in a stable manner. This power is the one which appears on the nameplates, specially in pre-2005 machines which is when the ISO 8528-1 regulation introduced power ratings for different operation modes. The figure below a 25 kW generator is shown:

Fig. I.4 Nominal power on the nameplate of a pre-2005 generator.

Prime Power (PRP) is the recommended power for the prime power mode. In this mode, the generator takes on a variable load for an unlimited number of hours. For example, an off grid hospital will require different levels of power (or 'loads') through the day and night. In this mode, the generator can be loaded up to 100% of the prime power value however the 24-hour average must not exceed 70% of this value. You can assume that the nameplate power of the pre-2005 generators and the prime power of the new ones are approximately the same.

Standby (or Backup) Power is the power recommended for operation in Emergency Standby mode, a mode that is limited to a maximum of 200 hours of operation per year. For example, the hospital is connected to the grid and the generator is there only as a backup for the occasional blackout. The generators used in development projects are generally run for much longer than this 200 hour limit.

Continuous Power is for applications where the generator runs at constant load all the time to avoid penalties for excess consumption, combined heat and power (CHP) and other economic applications that are less frequent in development projects. Do not get confused by thinking that the generator will be used non-stop with continuous power; this would lead you to choose a generator that is far too big! The difference is actually whether the load is variable or not.

In most cases you will be using Prime Power. Remember that *the same machine* has several power ratings according to the type of use it will be given. You can see this in the brochure below where the manufacturer of the DGBC generator rates its machine to be able to supply 44 kVA if used at variable loads over 200h per year (Prime power use) or stretched to supply 50 kVA if only used occasionally (Standby power).

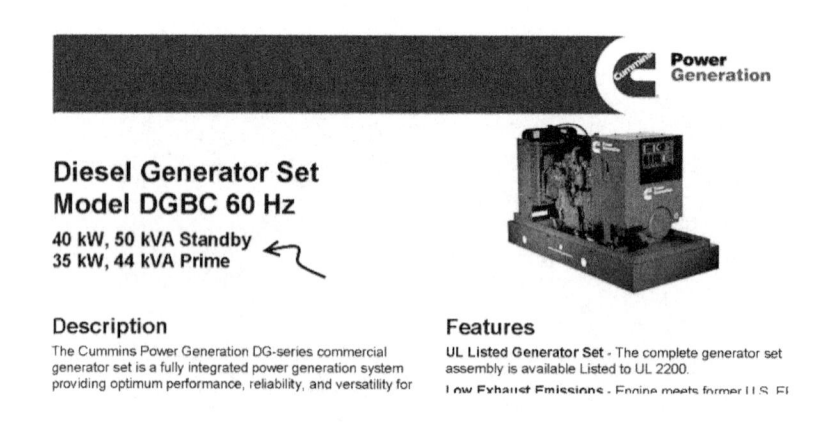

Fig. I.5 Standby and Prime Power specifications in a brochure.

Some basics of electricity

This section will help you to understand some basic electricity terms that frequently come up around generators. Don't worry if you don't understand it all at first.

The simplest form of electricity is direct current (DC) in which there is a difference in voltage or in electric potential between two points that causes the electricity to always circulate in the same direction.

The water analogy and units

An electric current is often compared with a stream of water. For water to start flowing a difference in height is needed. The steeper the slope from this height difference, the greater the propensity to flow.

Voltage is similar to this "height difference" or, in electrical terms, the difference in electric potential between the points. It is measured in **Volts (V)**, with the most common voltage being 220 V.

Current is the "flow rate" of electricity circulating. It is measured in **amperes (A)**.

Resistance measures the resistance of the electricity to flow in a material. It is measured in **Ohms (Ω)**.

Power is the amount of energy being used at a given time. The unit for this is the **watt (W)**. As the watt represents a very small amount of power, it is more common to see the term **kilowatt (kW)**, which is 1000 watts, being used.

Consumption is measured in kilowatt-hours (kWh). Or in other words, the consumption of a 1 kW appliance that has been running for an hour is one kWh.

Alternating current (AC)

In alternating current, the current does not always flow in the same direction as occurs with a direct current but instead varies cyclically in value and direction creating a sine wave pattern. The **frequency** measures how many times per second the current circulates in a given direction. Most countries use 50 Hertz (Hz) frequency.

14 Generators in development projects.

Given that the voltage constantly varies in AC following the cycle, it is necessary to define a representative value that is practical to use. An **effective value** or RMS value is thus defined as the value of a continuous current which would dissipate the same amount of power. Effective values are most commonly used.

So while 220 V is the effective voltage, the instantaneous voltage changes continuously between +311 V and -311 V depending on its momentary location in the cycle at a particular moment.

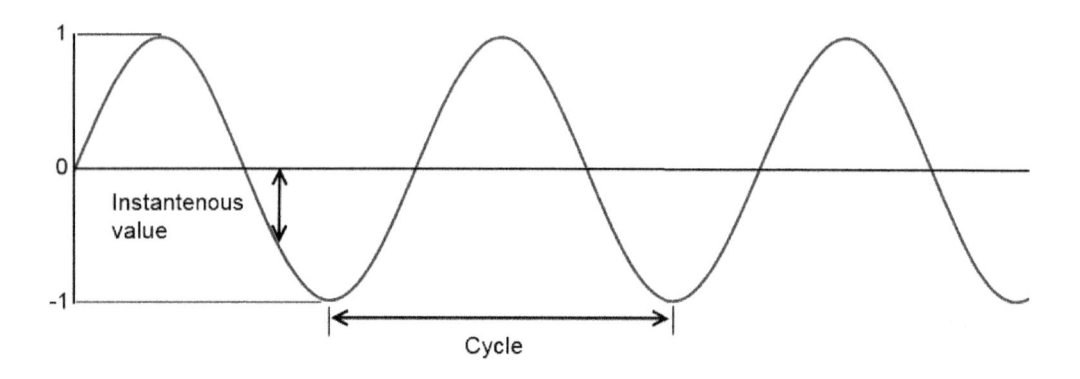

Fig. I.6 Alternating current sine wave

The main advantages of alternating current are that you can easily switch from one voltage to another using a transformer, motors and generators are cheaper and more efficient and it avoids undesirable effects such as electrolytic corrosion, magnetisation, etc.

In alternating current there are two elements that consume energy in a special way: the capacitor that stores energy in an electric field and the coil that stores energy in an electromagnetic field. This consumed energy, which is not really useful to perform work, is called **reactive power**, while the other type that does the work is called **active power** or real power. **Apparent power** is the sum of the two. Its unit is the kilovolt-ampere (kVA).

To understand the different powers in a more intuitive way, imagine a horse pulling a cart on a train track. If the horse pulls at an angle to the railway track, only part of its effort, the real power, will be used to move the cart along the rails. The reactive power is perpendicular to the tracks and does not contribute to the advancement of the cart. So, although the horse is apparently making a lot of effort (kVA), only the real power (kW) produces "useful" labour.

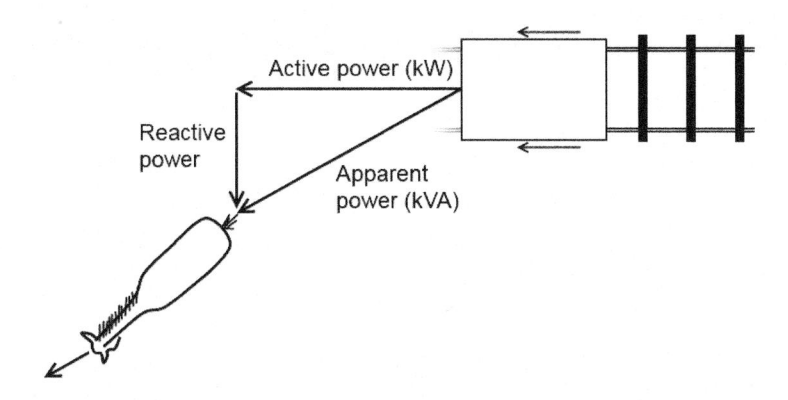

Fig. I.7 Horse analogy for the apparent, real and reactive power.

The relationship between these two powers determines the **power factor**, which measures to what extent the power is really being put to good use. It is usually close to 0.8 as a whole for a network (i.e. 0.8 kW = 1 kVA), although for each device it differs; it could be said that "*some horses are more clever about how they pull the cart*".

Three-phase systems

In three phase systems there are three alternating currents of the same frequency (called phases) that have a certain delay between them of around 120°, which are carried in a particular order. The rotation of the generator with 3 poles 120° apart is what creates three phase waves:

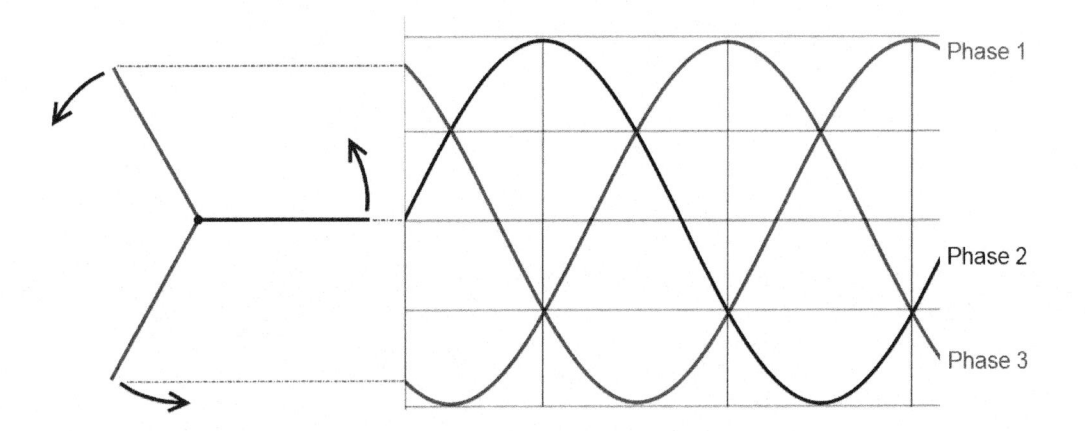

Fig. I.8 Three poles 120° apart that rotate create three phases in three-phase current.

16 Generators in development projects.

Most generators are three-phase. These can be connected directly to three-phase appliances and/or specific appliances may be independently connected to each of the phases.

Whereas in single phase system there is only one voltage, in three-phase systems there are two voltages depending on which terminals are taken for the measurement:

- **Phase voltage** is the voltage between each phase and the neutral.
- **Line voltage** is that which is measured between any two phases.

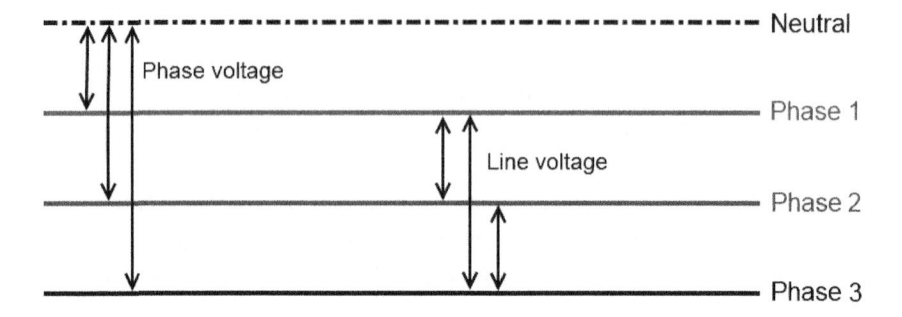

Fig. I.9 Line voltage and phase voltage in three-phase systems with 4 cables.

When a voltage is specified as 220V / 380V, the first figure represents the phase voltage and the second represents the line voltage.

Star or delta connection.

While connecting the three cables, all the cables may be attached at one end to form a **wye or star connection** or the end of each cable in the phase may be joined to the next one to form a **delta connection**.

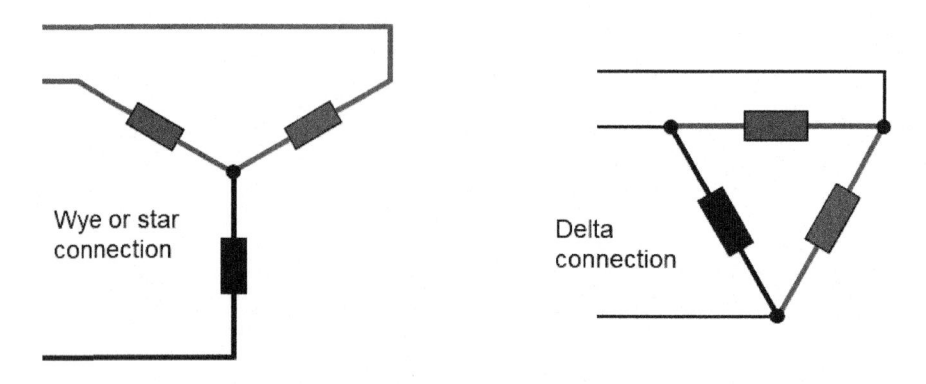

Fig. I.10 Wye connection and delta connection in three-phase systems.

In the star connection the line current is equal to that of the phase and the line voltage is $\sqrt{3}$ times (≈1,732 times) the phase voltage. In the delta connection the line and the phase voltage are the same.

The cables can be connected in four different ways to obtain power and voltage combinations: star-star, star-delta, delta-star and delta-delta.

The three-phase systems may have three cables or four cables (three of phases and the neutral).

Relationship between voltage, current and power for alternating current

While for direct current this relationship is simple (Power = Voltage * Current), three-phase alternating current is slightly more complicated because it changes depending on where it is measured, how many conductors there are and whether or not the system is balanced. To keep this introduction simple, Annex A explains the different rules and formulas for calculating power.

Working principle of an alternator

To produce electricity the alternator is based on the phenomenon that a moving magnetic field induces currents in a nearby conductor. The alternator has two fundamental parts:

- The **rotor,** which is the source of magnetism and rotates to provide the moving magnetic field.

- The **stator** which is fixed and has wire wound in the form of copper coils to receive the induced current.

Applying force to the rotor will make it rotate, as is done by a diesel engine in a generator. The movement of the magnetic field induces currents in the stator, producing electricity.

The source of magnetism are the electromagnets that are magnetised when they receive the current from a small DC generator called the **exciter**.

18   Generators in development projects.

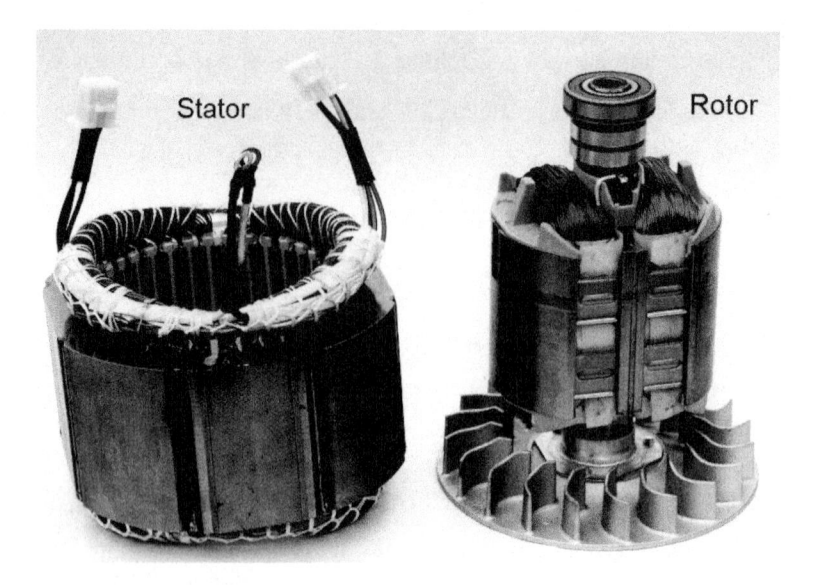

Fig. I.11 **Rotor and stator in a three-phase generator** *(Courtesy of Senci).*

As the voltage and frequency depend on the rotational speed, is very important to keep this speed within consistent range, and to always operate the diesel engine at the same speed. A consistent speed is achieved through the use of a **speed regulator,** which automatically accelerates or decelerates based on the engine load.

Generators also incorporate **automatic voltage regulators (AVR)** to maintain the voltage within a consistent range, regardless of the speed of the generator. Automative voltage regulators achieve this consistency by electronically adjusting the magnetic field within the rotor.

1

Economic considerations

Generators are one of the key components in the success of a project given the costs their purchase and operation represent.

An improperly sized or poorly designed generator can quickly ruin all other project efforts, and can result in a project being unsustainable for the communities it should serve.

Understanding the economic dynamics is essential before deciding what type of installation is needed. Do not skip this section!

Ballpark figures

Each generator and situation in which it operates is unique. Even so, to help with the planning, the section below sets out a series of numbers based on real situations that will give you an idea of what order of magnitude to expect.

How much could the generator cost?

While waiting for answers from suppliers about the price of the particular configuration you need, you can use these formulas to estimate approximate prices without delivery (based on 2013 prices):

Three-phase generator: $\quad$ Cost USD (2013) = $\dfrac{\text{Power kW} + 47}{0.0047}$

Single-phase generator: $\quad$ Cost USD (2013) = $\dfrac{\text{Power kW} + 30}{0.003}$

For example, if I need a 50 kW three-phase generator: $\dfrac{50 + 47}{0.0047}$ = 20,638 USD

Are there any economies of scale?

Larger generators consume less diesel per kWh produced, although there is little difference. For example, a 200 kW generator consumes 85% per kWh of that which a 20 kW one would.

The larger the generator, the lower the cost of power. This trend is very marked. If it is possible for the particular case you have at hand, it is cheaper to centralize the consumption using a single generator than to buy one for each consumer base.

For example, buying a 40 kW generator is 25% cheaper than buying two 20 kW generators.

How much diesel does it use?

Between 0.3 and 0.35 litres of diesel per kWh if it is correctly sized and maintained.

Diesel expenditure accounts for the overwhelming majority of the cost of a generator during its lifetime, as you can see in the figure, based on a price of $1 per litre.

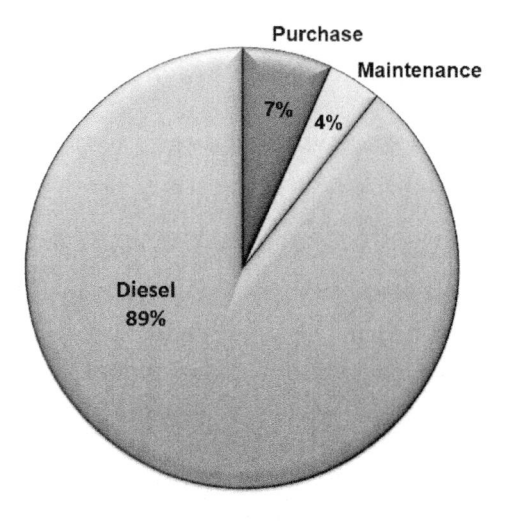

Fig. 1.1 Cost distribution for diesel at $1/litre.

If you run these numbers for a $20,000 generator, you will realize the importance of thinking about **how the costs will be dealt with**. It isn't enough to just buy the generator and leave. It is extremely important to also plan ahead and consider how the diesel to operate the generator will also be funded!

Generator	$20.000
Maintenance	$11.500
Diesel	**$254.000!**

What are the generator's maintenance costs excluding diesel?

Maintenance costs make up approximately 60% of the costs of the generator during its lifetime. The main outlay corresponds to reconditioning the generator at the halfway point of its lifespan, corresponding to approximately 30% of the initial costs of the generator.

The effect of load

One of the fundamental issues when making economic calculations, is the load level at which the generator runs. If a pump that consumes 40 kW is connected to a 100 kW generator, the load is 40%.

A generator that works with a very low load is very inefficient and reduces its useful life due to hard carbon deposition on the cylinders and the valves (cylinder glazing). When the generartor approaches its maximum load the efficiency increases somewhat but the general wear results in the generator wearing out much more quickly and not lasting as long.

Effect of the load on diesel consumption

Look at the graph below. Working at 8% load gives a 12% efficiency and working at 80% load increases the efficiency to 32%. The difference in performance results in the generator using three times more diesel per kWh for a load of 8% than for an 80% load.

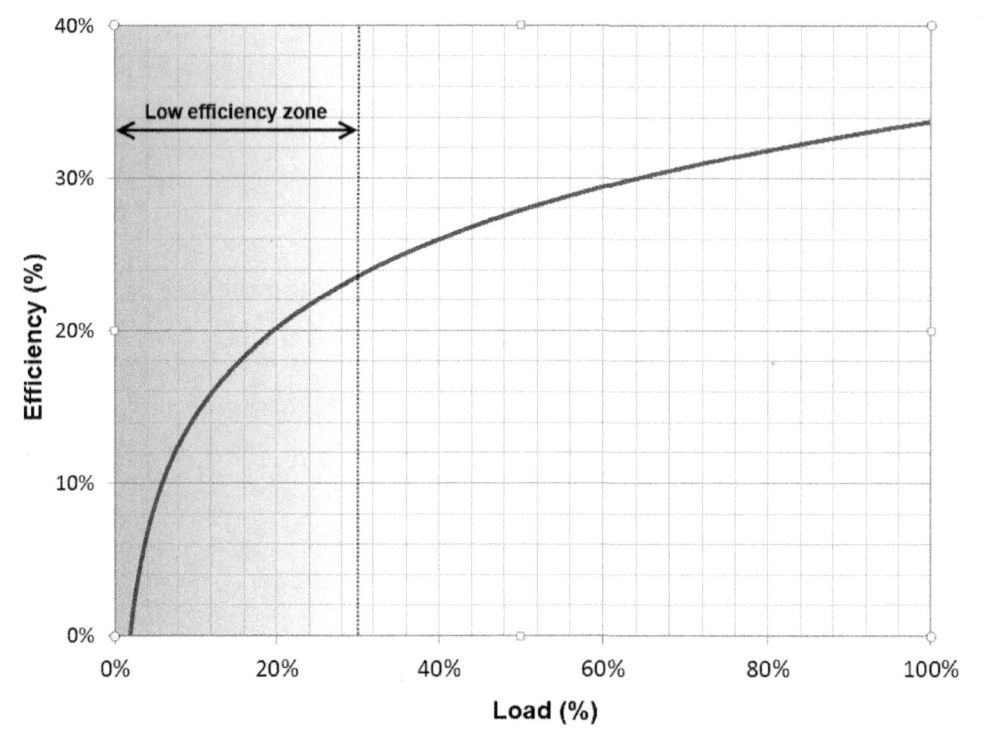

Fig. 1.2 Efficiency in the use of diesel according to load *(Source: Graph prepared based on consumptions obtained from: www.dieselserviceandsupply.com)*

Try not to install or use generators when the load will be less than 30%. If you were planning to use a generator the size of a bear to supply something as small as a laptop, consider alternatives like having a battery bank or a smaller generator for this type of use.

A properly sized, well-maintained generator generally consumes between **0.30 and 0.35 litres of diesel per kWh produced**.

Effect of the load on the lifespan of the generator.

Generators in acceptable working conditions and with proper maintenance generally last between 20,000 and 30,000 hours with an overhaul between 10,000 and 15,000 hours.

Never run a generator at more than 110% of its nominal power, not even for a brief moment. Overloaded generators deteriorate very quickly.

In this case the trend is reversed; the lower the load up to a point, the longer the generator will last.

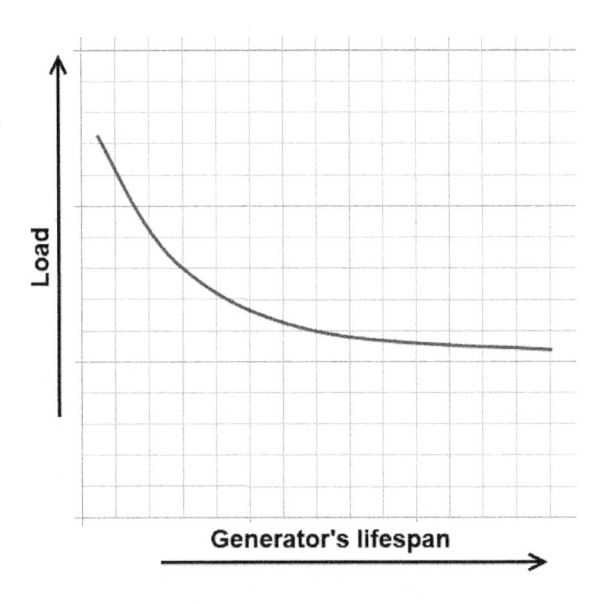

Fig. 1.3 Curve of the typical lifespan vs. load.

24 Generators in development projects.

As well as the value of the load itself, the average life of the generator is greater when loads are less variable.

To reconcile the two opposite trends seen above, ISO 8525-1 recommends that **the average load over 24 hours does not exceed 70% of the specified capacity of the generator** on the nameplate, for the majority of applications.

To calculate this average, only the times where the generator has been in operation are used and all the loads that are less than 30% count as 30%. For example, a load of 12% would still be counted as 30%, while a load of 31% would be counted as 31%.

What is the average load of a generator that has been operating over the last 24 hours, according to the following table?

Hours	Load
3	45%
6	80%
4	56%
2	18%

The last two hour period is taken as 30%, the load having been below 30%.

$$Al = \frac{3*45 + 6*80 + 4*55 + 2*30}{3+6+4+2} = 60\%$$

Estimating consumption

Although consumption varies depending on the circumstances, you can use the graph below to anticipate what your generator's consumption is likely to be, or determine the performance of an existing generator.

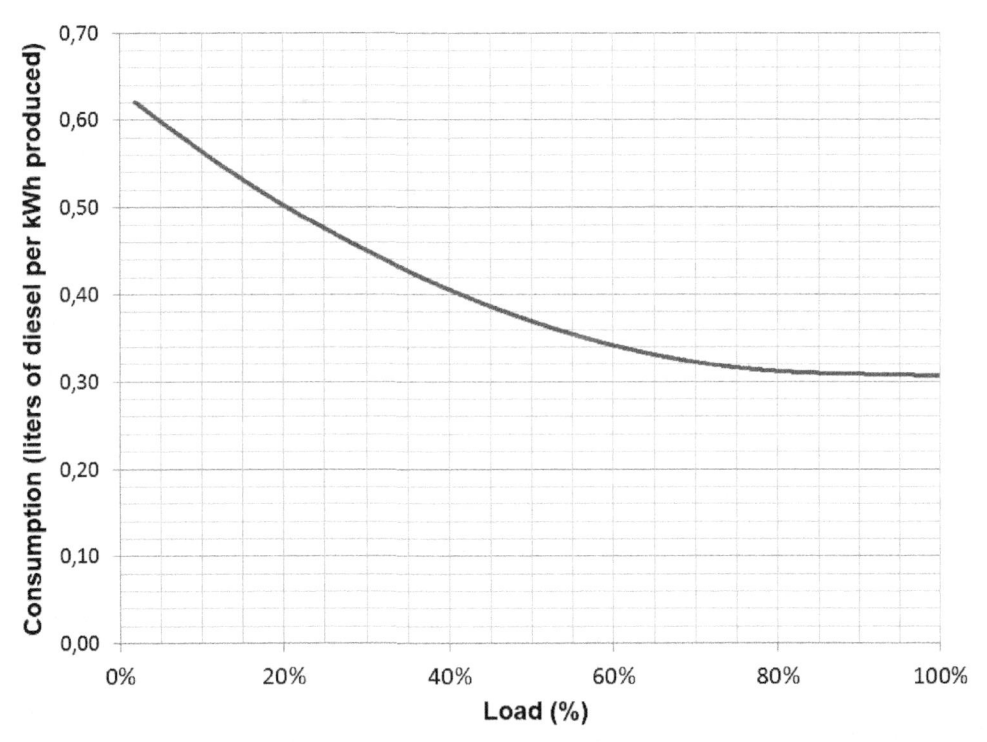

Fig. 1.4 Diesel consumption per kWh produced *(Source: Graph prepared based on consumptions obtained from: www.dieselserviceandsupply.com)*

Approximately how many litres of diesel has an 80 kW generator consumed after 12 hours of work with an average load of 40%?

In 12 hours at 40%, an 80 kW generator produces:

Energy = time * load * power = 12h * 0.40 * 80 kW = 384 kWh

Looking at the graph above, a 40% load corresponds to 0.408 l/kWh, thus:

C = 384 kWh * 0.408 l/kWh = 156.67 litres of diesel

Trends in the price of diesel

In the last 20 years (1989-2009), the price of diesel has tripled.

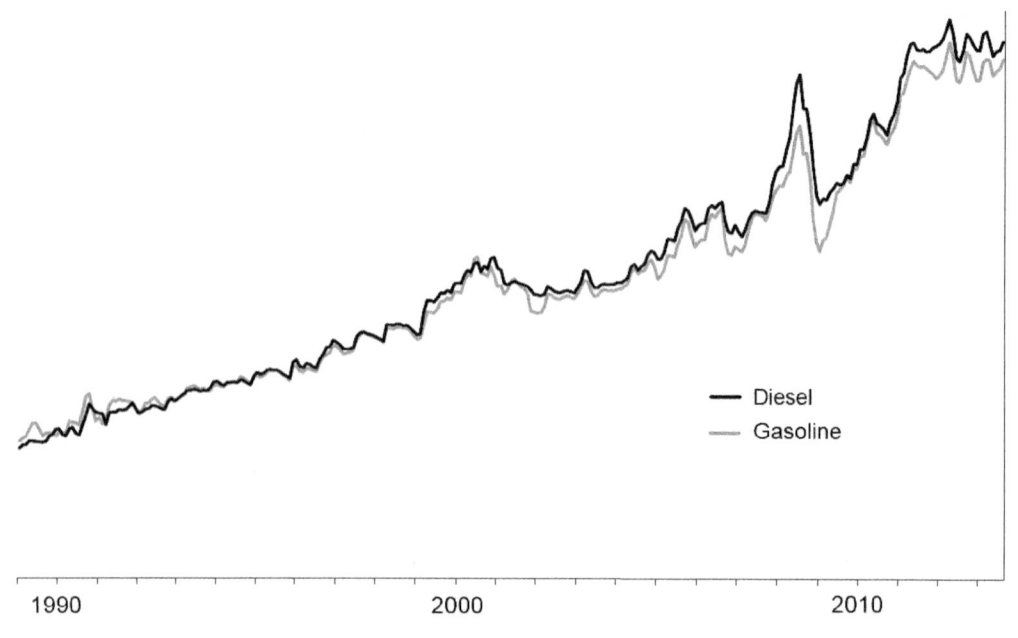

Fig. 1.5 Evolution of the price of diesel and gasoline *(Source: Quarterly Energy Prices DEEC)*.

Although there are many throrough reports on future trends, the final mesage is that you can basically expect almost anything, although the centre of the forecasted range below indicates that prices are likely to continue rising over time.

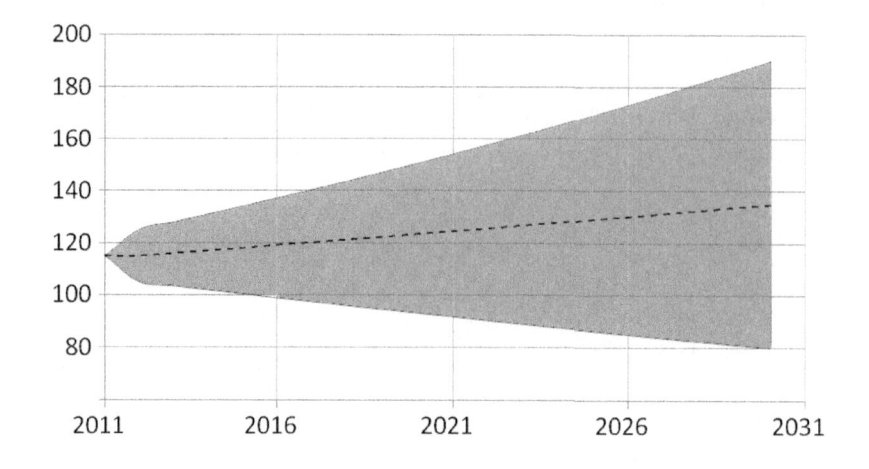

Fig. 1.6 Projected evolution of the price of a barrel of petroleum *(Source: DEEC)*.

As the trends are quite unpredictable, make a **contingency plan** as to what measures can be taken to make do in case of a major rise in the price of diesel.

Fig. 1.7. Without fuel during the Nazi invasion of Holland, 1940, *(Nationaal Archief)*.

Moreover, the fluxuation of the price of diesel in your area may have large seasonal variations, due to increased demand for heating in winter, hampered navigability of the sea, transport problems with the monsoon, etc. In some cases it may just not be readily available.

Is it worth investing in low consumption appliances?

Within your development project, you will often have to choose between investing in an expensive 'low consumption' appliance that consumes less power, or a cheaper appliance that consumes more power.

A simple way to help make this choice is to compare the expense incurred in buying and operating each alternative. To do this we must consider two things:

- You have to know the **lifespan** of the appliance or make an informed assumption as to how long it will be working. For example, a lightbulb will last three years on average based on its usage or a new health care facility will be designed to last for a period of 30 years.

- **The value of money decreases over time.** A cup of coffee that costed 60 cents in the year 2000 now costs $1.50. One dollar now buys more than one dollar will buy in 20 years time. In order to compare invoices they must be from the same moment in time, usually from when the project first starts.

Figuring out the annual bill of an investment

This is about figuring out the yearly equivalent cost over the lifespan of an expenditure that was made in one payment at the beginning. In other words, if I spend $20,000 on a generator that will last five years, what is the equivalent amount if I paid annually?

The procedure is as follows:

1. Find out the **interest (i)** that a bank would give you for a deposit of a similar amount and transform it to a decimal. For example, for 3%, i = 0.03.

2. Take an informed guess as to what the **inflation (s)** may be in your country for the period of time you are considering. You can look at some of the years in the World Bank data and give it your best estimate: http://data.worldbank.org/indicator/FP.CPI.TOTL.ZG/countries/all?display=graph. We say guess, because it is impossible to know how the inflation will evolve as time goes on. The inflation, is your parameter s, also as a decimal.

3. Calculate the **real interest rate (r)**. This rate takes into account the interest and inflation. If inflation is higher than a bank would give, the money is worth more in the present than it would be worth in the future. If they are equal it

maintains the money's value, and if the bank's interest is higher than inflation, the value of the money will increase with time. The real interest rate is calculated by using:

$$r = \frac{1+i}{1+s} - 1$$

4. Calculate the **amortization factor (a$_t$)** for T years:

$$a_t = \frac{(1+r)^T * r}{(1+r)^T - 1}$$

5. The annual bill for the investment is the amount invested, M, multiplied by the amortization factor:

$$F = M * a_t$$

Remember that the calculated amount is just an approximate value and that there are often other parameters that affect the final decisions. Therefore, do not get bogged down because the data has some uncertainty.

A telecommunications system that operates 24 hours a day is going to be connected to an existing generator that burns 0.33 litres per kWH produced(l/kWh). The load of the system is low in relation to the total of the generator and will not affect its yield. The bank interest is at 2%, inflation at 4% and the price of diesel is $1/litre. These are the two options available:

	Appliance A	Appliance B
Cost ($)	7.000	13.000
Consumption (W)	800	240
Duration (years)	10	15

We are going to calculate the investment and running costs for the two options and see which outcome is more economic. Both systems will be operating 365 days per year, or:

T= 365 days/year * 24 hours/day = 8,760 hours/year.

APPLIANCE A:

Its annual consumption will be 8,760 hours * 800 watts = 7,008,000 Wh = 7,008 kWh

30 Generators in development projects.

The annual bill for this consumption is: 7,008 kWh * 0.33 l/kWh * $1/l = $2,312.64

The calculation of the annual bill for the investment is as follows:

The real interest rate is: $r = \dfrac{1+i}{1+s} - 1 = \dfrac{1+0.02}{1+0.04} - 1 = -0.0192$

The amortization factor is:

$$a_t = \dfrac{(1+r)^T * r}{(1+r)^T - 1} = \dfrac{(1+-0.0192)^{10} * -0.0192}{(1-0.0192)^{10} - 1} = 0.08973$$

The annual bill: F = M * at = 7,000 * 0.08973 = $628.11

The total annual cost of this option is the sum of both bills:

$2,312.64 + $628 = $2,940.75

APPLIANCE B:

Its annual consumption will be 8,760 hours * 240 watts = 2,102,400 = 2,102.4 kWh

The annual bill for this consumption is: 2,102.4 kWh * 0.33 l/kWh * $1/l = $693.8

The calculation of the annual bill for the investment is as follows:

The real interest rate is the same as for the other appliance.

The amortization factor:

$$a_t = \dfrac{(1+r)^T * r}{(1+r)^T - 1} = \dfrac{(1+-0.0192)^{15} * -0.0192}{(1-0.0192)^{15} - 1} = 0.05687$$

The annual bill is: F = M * at = 13,000 * 0.05687 = $739.31

The total annual cost of this option is the sum of both bills:

$693.8 + $739 = $1,433.11

Therefore, investing in the more expensive appliance is well worthwhile because it results in $1,500 of savings per year.

Can the operating costs be covered?

Earlier in this book, we discussed a generator of 80 kW running at 40% for 12 hours a day. This is a fairly realistic example that would generate a daily consumption of 156 litres of diesel. 156 litres per day is easily translated into $80,000 per year and that ... is a lot of money for most organizations!

It is vital **to make sure that someone is thinking about whether the costs are manageable over the long term**. The following is an example of how such considerations can be addressed.

A plan is being discussed to install a pump and a generator to distribute well water in a community of 300 families with an average of six members per family. The average family income is $280 per month. The system would pump 2,000 litres of water per kWh consumed. The pump and the generator would be replaced every five years at a cost of $18,000 and the diesel costs $1 per litre. Would this plan be affordable for the families?

The United Nations recommends that families do not pay more than 3% of their income on water bills. This is the key; always seek important reference information like that to help you to better determine if what you are planning is reasonable or not.

The rest you already know:

If the planned allowance is 30 litres per person per day, the operating costs would be:

30 l/per.*day * 6 people/family * 300 families * 365 days/year = 19,710,000 l/year

$$= \frac{19{,}710{,}000 \text{ l/year}}{2000 \text{ l/KWh}} = 9{,}855 \text{ kWh}$$

Assuming an efficiency of 0.32 l/kWh, the annual operating bill would be:

9,855 kWh * 0.32 l/kWh * $1 / = $3,153.6

If expenditures on diesel accounts for 89% of the total costs during the lifespan of the generator and maintenance 4%, the maintenance cost will be:

$$\frac{4\% * \$3153.6}{89\%} = \$141.73$$

The total annual operating bill is therefore: $3,153.6 + $141.73 = $3,295.33

The calculation of the annual bill for the investment, assuming for this example an interest rate at 1% and inflation at 6%, is as follows:

The real interest rate is:

$$r = \frac{1+i}{1+s} - 1 = -0.0472$$

The amortization factor:

$$at = \frac{(1+r)^T * r}{(1+r)^T - 1} = \frac{(1+-0.0472)^5 * -0.0472}{(1-0.0472)^5 - 1} = 0.1726$$

The annual bill: F = M * at = $18,000 * 0.1726 = $3,106.8

The total annual cost of this option is the sum of both bills:

$3,295.33 + $3,106.8 = $6,402.13

The share per family: $6,402 / 300 families = $21.34

Each family earns: $280 /month * 12 months/year = $3,360 per year.

Thus expenditure on water represents: $\frac{21.34}{3,360} * 100 = 0.64\%$ of their income

An expenditure of 0.64% of a family's income is far less than the United Nation's maximum recommended expenditure of 3%. So this would appear to be a very good investment.

Presumably the system is economically viable **if people are willing to pay**. The willingness to pay must be investigated as part of the overall project.

Prioritizing the loads

Bases on what you have read so far, you've probably come to the conclusion that running a generator is expensive and that it is worth thinking about what to connect and what not to connect.

Left up in the air, people go around connecting everything to the generator until it can take no more. This causes the system to operate outside the range where it is efficient, to rapidly deteriorate and to have things that add very little value to the community connected to a very expensive system. An economic disaster! **It is essential to manage what is connected to a generator and their order of importance.**

Planning the load: Creating a list of priorities and timeslots

Planning the load simply involves listing the relative importance of the various loads and defining the times at which they can operate. For example, lighting the perimeter of the employee compound of a refugee camp is critical to safety and occurs at specific hours during the night. Pumping water for use in the camp is also of utmost importance but, as there is a reservoir supply, it can be done at whatever time of the day is most convenient.

All of this can be organized in a diagram similar to the one below:

Load	kW	Priority	Time of day (1–24)
Security lighting	3.5	1	1–6, 19–24
Water pumping	7	1	
Charging telecoms eq.	0.2	1	
Interior lighting	6	2	
Workshop	5	2	
Washing machine	2.3	3	
...	...	...	
Watching cat videos	0.75	999999	?! ?! ?!

Fig. 1.8. Simplified example of a priority list of loads and timeslots.

If you complete the table based on the load for each hour and optimize it as you go, you will end up with the best usage. Don't create too many priority levels that will later complicate your life; with two or three you will probably have enough.

Consolidating the load

Once created, the table of priorities and timeslots must be consolidated to help you determine the smallest possible generator that runs for the fewest hours possible at a load as close as can be to 70%, all while reasonably considering the needs of the users. Don't let the generator end up scheduling births in the maternity ward!

You can think of it like a game of Tetris; it is about fitting in the different blocks of consumptions, especially those that allow flexibility, so that everything can be run with the smallest machine possible that operates in its most efficient range. Most frequently it can easily be done and it is only a matter of organization.

Creating differents circuits

It is one thing to create a beautiful timetable, but another thing entirely to ensure that it is adhered to. **There must be an ability to control what is connected to each line.**

> The Mtabila refugee camp generator works erratically between 12:00 to 14:00 each afternoon, shutting down every 5 or 10 minutes. It turns out that employees are connecting electric cookers to modified lightbulb sockets in their homes until the generator conks out. When they lose electricity, they have learned to disconnect the cooker for a few minutes and then later try to reconnect it before the food gets cold, perpetuating a steaming cycle of shutdowns and start-ups of the generator.

Apart from all the economic problems and the wear on the generator which you already know about, another major problem in such an scenario is that **the reliability of supply is lost**. Imagine a vaccine refrigerator running under this regime!

If circuits are separated according to their priority, thay can be selectively turned off as the generator exits its operating range, thus avoiding situations such as the one just described.

Rather than blowing a fuse over such unauthorized connections, take into account that these workers **are expressing a need** and that cutting off the supply would only solve your part of the problem. Perhaps you could instead arrange the delivery of gas cylinders or a common dining area to solve everyone's problems.

Small, constant and important consumptions

Imagine a small NGO office in a remote location with a large 60 kW generator installed for various uses. They are not going to keep that size generator running for eight hours just to supply a few 40 W computers and the coffee machine!

Diesel generator and battery hybrid system

One solution when you have a large generator but a small power requirement at times, is to add a battery bank and an inverter charger that converts the direct current into alternating current, and vice versa, while also acting as a charging unit. If the consumption is made in DC, only a charger is needed which is far less expensive than an invertor. When the generator is turned on for other reasons, the small of charging the batteries is added to the normal load. Once the generator is turned off, the small consumptions are fed by the batteries until the next cycle of generator use.

It is important to note that this option on its own is not very viable at the time this book was written, since the cost of producing and storing energy is greater than that of using a small additional generator. Over the next few years the cost of storing energy in a battery is expected to fall, and this option will become an increasingly viable and low cost solution.

What is the cost of storage per kWh of 4 Trojan T-105 6 V batteries and C_{20}=225 Ah with a total cost of $800 which are going to be used with a 50% depth of discharge?

The energy that is discharged during each cycle is:

$$E = C_{20} * V * DoD * 4 \text{ units} = 225 \text{ Ah} * 6 \text{ V} * 0.50 * 4 = 2{,}700 \text{ Wh} = 2.7 \text{ kWh}$$

According to the manufacturer's information, with a 50% discharge the batteries last for 1,500 cycles.

The battery cost per kWh stored is:

$$\text{Storage cost} = \frac{\$800}{2.7 kWh/cycle * 1500 cycle} = 0.1975 \text{ \$/kWh}$$

In addition to storing each kWh, the kWh must also to be produced. In the process of charging the battery, generally 25% of the energy is lost. If it is produced with a generator that consumes 0.34 l/kWh, a litre of diesel costs $1 and with maintenance not being taken into account:

Cost of production = 0.34 l/kWh * $1/l * 1.25= 0.425 $/kWh

The total cost would be: 0.425 $/kWh + 0.1975 $/kWh = 0.6225 $/kWh

A small additional generator of this capacity, even with all the maintenance included, is currently cheaper than the battery option. It also produces the electricity right when it is needed, without the need to take on the storage cost. Nevertheless, investing in battery storage may be an appropriate solution for other reasons.

Parallel system of batteries and solar panels

If solar panels are added to the battery system, it doesn't greatly increase the cost. Is in fact worthwhile economically, because the storage cost is much cheaper than burning diesel and the extra investment is very low. Therefore, if in the above example the batteries were charged using solar panels instead of charging them via the generator, you would be saving:

0.425 $/kWh – 0.1975 $/kWh = 0.2275 $/kWh

However, keep in mind that these systems pay back their value very slowly, often over a number of years. What if someone gets it wrong and adds well water instead of distilled water to the batteries, the panels are stolen or they end up being diverted to the headmaster's house? According to a study of 3,000 solar panel systems installed in third world countries, five years after the date of installation, one in four solar systems doesn't work.

Over five years of operation, the savings would be:

4 years * 365 days/year * 2.7 kWh/day * 0.2275 $/kWh = $896.8

Hybrid system

A further step in sophistication, complication and investment (but with huge savings in operating costs) is to launch a hybrid system where the solar panels or a wind turbine charge batteries. The batteries can then help the generator at peak load times and removing load in general.

Hybrid systems are considerably more expensive at first but much cheaper over the life of the system. Seek additional help if this is what you need.

In any system that uses batteries, consider the following:

If you are using sensitive equipment such as a computer, it is important that the invertor is **pure wave or sine wave**. The modified waves only approximate the wave and will damage sensitive equipment.

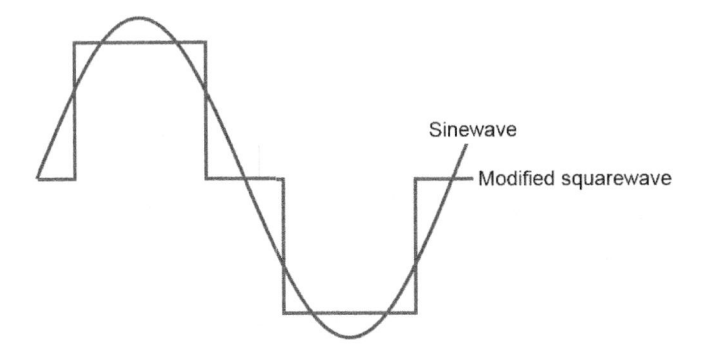

Fig. 1.9. Pure sinewave vs. modified squarewave.

Heat recovery units

Only a small part of the energy from a generator is transformed into electricity while the largest part of a generator's energy is in fact lost in the form of heat (exhaust gases, cooling circuit, heat radiation, etc). The energy distribution is roughly as follows:

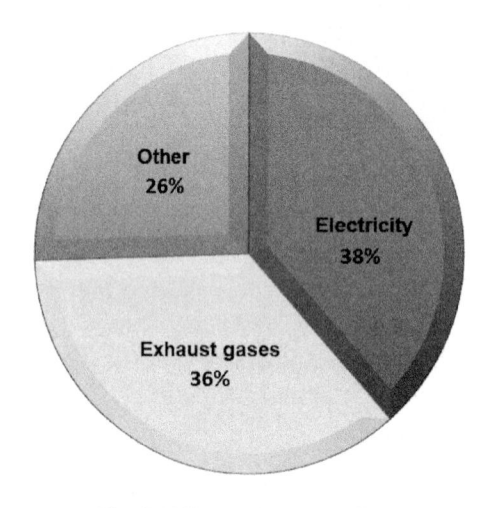

38% Electricity
36% Exhaust gases
6% Heat radiation
11% Cooling circuit
4,5% Turbocharger supply.

Fig. 1.10 Energy balance of a generator at full load *(Source: Mahon, 1992)*.

Through the exhaust, about the same amount of energy is lost as is used to make electricity!

In some cases it is worthwhile and appropriate to install a heat exchanger that recovers heat from the combustion gases and generates steam which can be used in other applications, for example for heating.

2

Sizing

In the exercises that follow, the sizing is done so that **the average power in 24 hours does not exceed 70%** of the nominal or prime power, and **at no time passes 100% of the power**. The allowable voltage and frequency drops are taken into account when specifying the class type G1, G2, G3 and G4. All this is explained in more detail below.

The focus in this book is to **adjust the sizing as much as possible** without extra safety margins because, in our experience, generators do not stop being used because at times they fall slightly short but because the costs they generate can not be met. And human nature being what it is, if a machine is capable of handling more load, it ends up being loaded with consumptions that are not as important, but do contribute to the final bill in full effect. In this context, erring on the side of safety is installing slightly small generators.

There are programs which help to size generators (for example GenSize from Cummings), although their sizing criteria may be quite different from those necessary in development contexts and may end up proposing solutions that are overly sophisticated and maladapted to your needs.

Finally, bear in mind that there may be local codes and standards in your country that affect the sizing of the generator and its installation. This is often the case for those generators that are intended for backup use.

Types of loads

Not all appliances consume power consistently or in the same way. The following are common types of load patterns:

- **Devices with constant consumption,** such as incandescent bulbs, kitchen resistors or electric radiators.

- **Devices with a peak in the startup** in which substantially more is consumed for a few seconds than that consumed thereafter. These are appliances that have motors. In these cases, the generator has to be able to cope with the sharp increase in consumption at the starting peak.

- **Devices with a variable load,** such as those using UPS or pulse technology battery chargers.

- **Devices with a very low power factor,** which for a low useful kW consume a lot of kVA. They typically require power factor correctors that also consume.

In linear loads (such as incandescent bulbs, heating or motors) the current depends on the voltage. In non-linear loads (almost any electronic device such as computers, printers, photocopiers or televisions) the current is independent of voltage. As a result, appliances with non-linear loads tend to be problematic because they introduce harmonics that alter the waveform and overheat the generator winding.

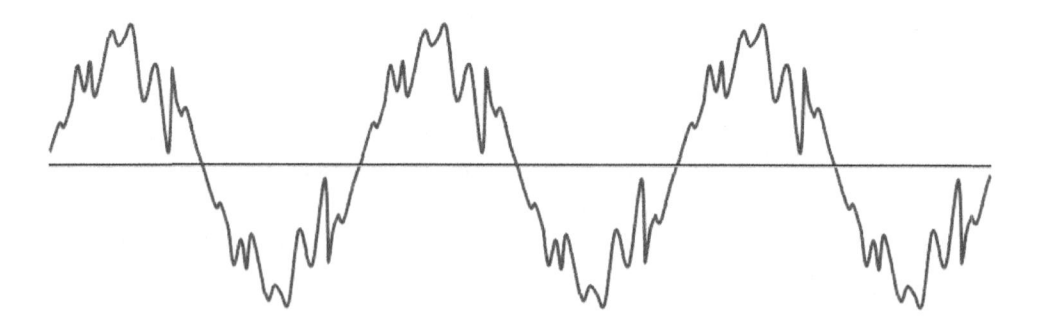

Fig. 2.1 Wave altered by harmonics.

Caution with some appliances

Some appliances are especially sensitive or problematic:

- **X-ray, MRI and CT machines** need voltage drops lower than 10% to avoid image defects.

- The **UPS** introduces a lot of harmonics noise. Make sure that the UPS load does not exceed 15%-20% of the generator for passives and ask the manufacturer about the double conversion type.

- **Motors with frequency variators** that introduce harmonics should not account for more than 50% of the load.

- **Battery chargers** operating on pulse can overload the generator.

- **Regenerative loads** in machinery. Cranes, lifts, hoists and certain other devices require the power supply to absorb energy during braking. A generator, especially when it is not loaded, will have difficulty absorbing this energy.

Estimating the power requirement

If there is an existing electric installation the best thing to do is to measure the consumption, as explained below. Otherwise you will have to find out the consumption of each appliance before scheduling them into a table, as you saw in the previous section.

Always measure if you can. Avoid surprises due to omissions, errors, misrepresentations or information that is too generic and has been taken from who knows where. Once the generator is installed, these surprises are expensive and complicated to fix. Do not use generic consumption tables that you find on the Internet. If you can not measure the consumption yourself, ask it to the manufacturer of the equipment that you are buying.

Understanding the significance of the power on the appliance nameplate

Almost all appliances come with a nameplate like the one below where the power is specified:

```
UL LISTED
HEATER
234R

MODEL No. QTV-22
120V AC  60Hz  12.5 AMPS  1500 WATTS
          MADE IN CHINA
LAKEWOOD ENG. & MFG. CO.
CHICAGO, ILLINOIS 60612
```

Fig. 2.2 Specifications label of an appliance.

The power is either directly shown on the nameplate, or can be easily calculated as the product of volts and amps. In this case, for example, the indicated power is 1,500 W, which corresponds to the product of 120 V * 12.5 A = 1,500 W.

When using the labels, take some precautions:

- Power consumption depends greatly on the way an appliance is used. Some appliances, such as workshop machinery, can consume more if they are used for demanding jobs.

- In other cases, the data printed on the label is not the power consumption but the load that the appliance accepts. For example, the starter on the right does not consume 65 W but is capable of starting up fluorescents up to 65 W.

- Some consumption can be hidden. Following the example of the fluorescents, on top of the starter fluorescents also have ballasts. Old ballasts can consume more than the tube itself.

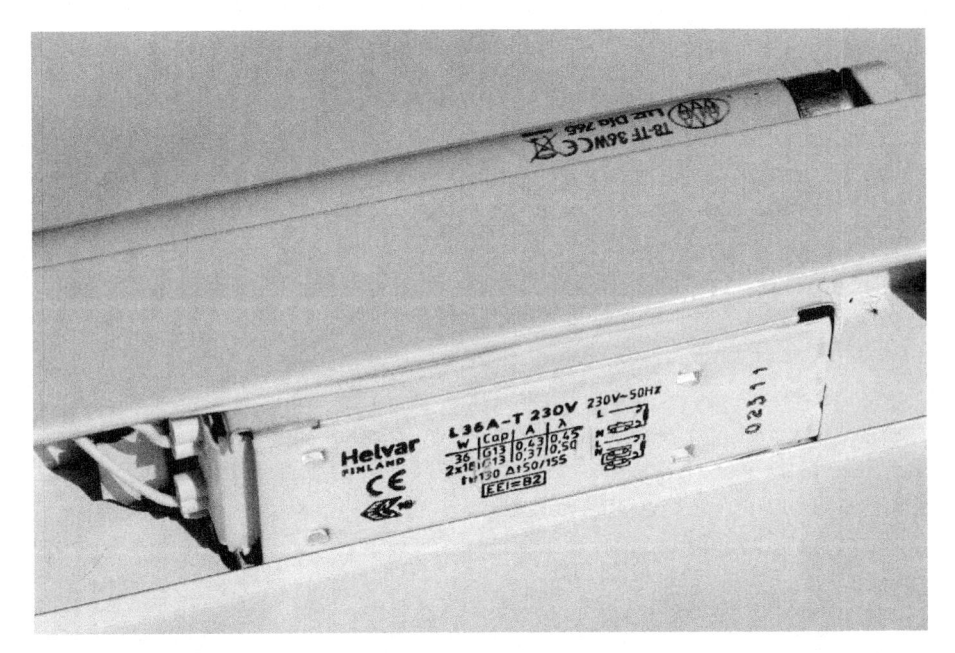

Fig. 2.3 Detail of a hidden ballast in a fluorescent light housing.

Remember, if you have the appliance on hand, measure its actual consumption! The above fluorescent was recorded at 110 W when measured despite only having a 36 W tube. Without having measured, you could have ordered a generator three times too small!

At the time of lighting with fluorescent lights, replace the ballasts of old fluorescents with electronic ballasts which consume far less. If you do not know which type of ballast you have, the old ballasts weigh much more, as though hefting a solid metal item.

Optimize your consumption before you size the generator. The generator and appliances are a whole package; there is little point operating the generator very efficiently with wasteful appliances.

Installation losses

Part of the current passing through a cable is lost through cable resistance causing a voltage drop. The longer the cable and smaller the section, the greater the voltage drop. Installation cables are sized so that the voltage drop does not exceed a certain percentage (usually between 3% and 5%).

However, these losses **do not affect the sizing of the generator** because, as the voltage is fixed in advance and can not increase to offset losses without overloading other devices on the same network, it is the appliance itself that has to make do with less electricity.

Peaks on starting motors

The easiest way to determine the peak is usually to ask the manufacturer. Annex C lists the peak values suggested by manufacturer Grundfos for one of the most common motors, those of submersible water pumps.

Often, motors have a letter assigned according to the current in kVA consumed by the motor with the rotor locked according to the NEMA code for motors. To calculate the start-up power in kVA, multiply the horsepower by the corresponding factor from the table below (average values):

Letter	Factor	Letter	Factor	Letter	Factor	Letter	Factor
A	2	F	5,3	L	9,5	S	16
B	3,3	G	5,9	M	10,6	T	19
C	3,8	H	6,7	N	11,8	U	21,2
D	4,2	J	7,5	P	13,2	V	23
E	4,7	K	8,5	R	15		

Fig. 2.4 NEMA motor code letter and multiplication factor for start-up kVA.

For example, if the nameplate on the motor has the letter A and it is 10 HP, the start-up current will be approximately:

$$10 \text{ HP} * 2 \text{ kVA/HP} = 20 \text{ kVA}$$

Measuring the consumption of a particular device

The easiest and cheapest way to measure how much a particular device consumes without having to open the circuit, is by using a current clamp or a power clamp meter. These tools measure the magnetic field produced by the electricity flowing through the wire, by surrounding it with hollow tweezers.

In the case of the current clamp, a reading is obtained in amps and from this the voltage is calculated. The monitor displays the calculation and allows you to record the consumption over a long period of time.

Fig. 2.5 Catalogue image of a current clamp Kuwell KW-266.

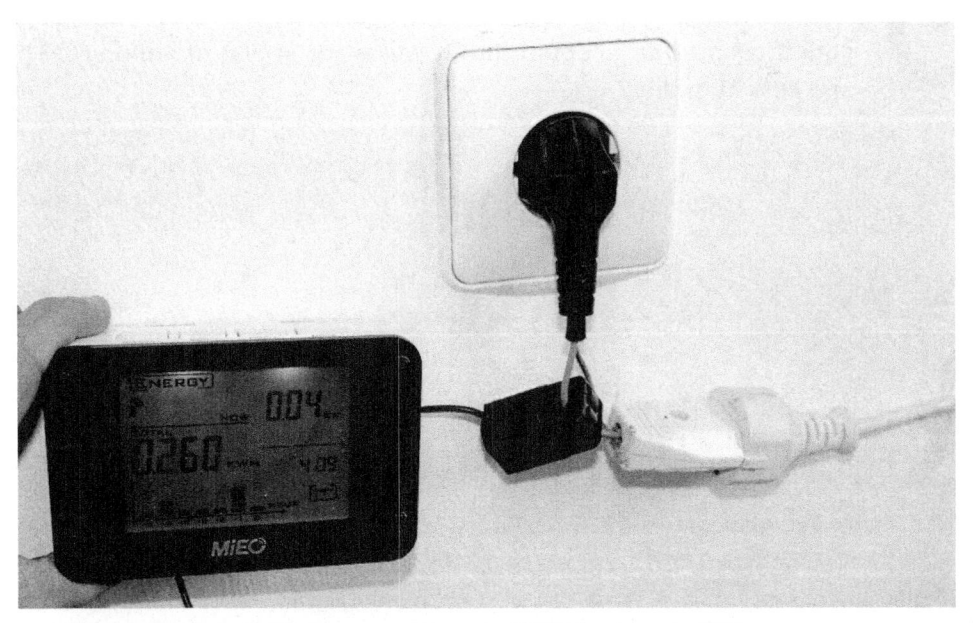

Fig. 2.6 Measuring with a power meter up to 17 kW (costs under $50).

When measuring with either, it is very important to enclose only one wire, otherwise what you get is the vector sum of several wires, which often is zero. To avoid separating the cables to expose the wires, you can use a small extension with the wires separated which is plugged into the appliance to be measured.

Derating according to the operating conditions

Much as with people, when it's hot, there is a lack of air and when there is too much moisture or dust in the air, generators lose their capacity to do work. To compensate for the power loss in these environments the generator is **derated.** Use the manufacturer's data to see how much to derate. Some common values are:

- For **high temperature,** increase the power by 0.2% per degree centigrade above 25 °C.

- For **high altitude,** increase by 1% for every 100 m above 1000 m altitude.

- For **high humidity**, 1.5% for every 10% of relative humidity above 30%.

- For **dusty conditions**, adjustments depend on the air inlet restriction caused by the additional air filters needed.

If a 100 kW generator is planned, what extra power will have to be added taking into account that it will work at 500 m altitude, 36 °C and 40% humidity?

Altitude: $\dfrac{500m - 100m}{100m} * 1\% = 4\%$

Temperature: 36 °C – 25 °C * 0.2% = 2.2%

Humidity: $\dfrac{40\% - 30\%}{10\%} * 1.5\% = 1.5\%$

The total derating is: 100 kW (0.04+0.022+0.015) = 7.7 kW

Therefore, a generator of at least 107.7 kw must be considered.

Several generators in parallel or a single generator?

Although you can operate several generators in parallel, the reality is that this is neither practical nor economical for sizes smaller than 500 kW, such power being very uncommon in the context of development projects. As such, operating multiple generators in parallel is usually not a good idea.

To operate two or more generators in parallel, or a generator and another alternating current source (e.g. mains electricity, a wind generator or a micro-hydroelectric plant), they must have the same frequency and be in phase. This is very difficult to achieve and requires special equipment.

Fig. 2.7. Eight Pramac GSW generators in a parallel install, Atacama Desert, Chile.

Size your generator with care and think about future consumption because the size decision is not easily reversable once the generator has been purchased. You can not just buy a 20 kW generator to supplement the existing 50 kW generator that fell short.

Rather than having two generators in paralell, it may sometimes be a good idea to have two very different size generators operating alternatively, one for the large consumption activities and the other for small consumption activities.

Voltage and frequency variations with load

Ideally, the motor of a generator should rotate at a constant speed, which is what determines the frequency of the supply. For example, rotating at 1,500 rpm gives a frequency of 50 Hz. In practice, this frequency varies within a small range until a load is applied.

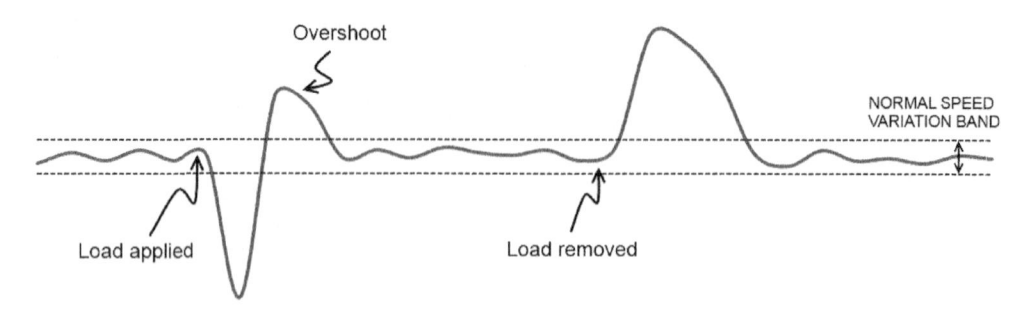

Fig. 2.8 Variation in speed in response to the application of significant loads.

The application of a load causes the generator to slow down. When the generator detects that it is being slowed, it increases the fuel input and accelerates again to try to regain cruising speed. In this process there is generally a slight overcompensation which causes the generator to go faster than needed before slowing down shortly thereafter. After stabilization, if the load is switched off, the generator accelerates until it corrects itself again by closing the fuel inlet. These speed variations are also accompanied by variations in voltage.

These variations, which are almost unnoticeable in an incandescent lightbulb, can make bones mysteriously disappear in radiographs or ruin a dozen computers. In addition, motors lose power with the cycle of the drop. If the voltage drops to 70%, the power falls to 49%.

So, when ordering a generator, in addition to the power, you should also specify the **type of response** as it has a big impact on the size and construction of a generator. Type of response is explained later in detail in the purchase options chapter.

If you specify a response type (ISO G1-G4) or check in the specifications that the frequency and voltage variations are within the tolerance of your equipment, you do not have to oversize more, as is sometimes recommended.

Sizing for constant consumption

What size of generator is needed for the night lighting of a compound consisting of 100 fluorescent tubes of 36 W and 70 PL bulbs of 20 W if the operation conditions at night are 30° C, 800 m above sea level and 40% relative humidity?

It is rare that a generator is only used for one thing and also for it to have a constant consumption, but this simple case serves as a good conceptual exercise.

1. We found out that the fluorescent units have a 20 W ballast so that total consumption is:

    P = (36 W + 20 W) * 100 units + 20 W * 70 units = 7,000 W = 7 kW

2. We want the generator to work at around 70% so as to have a good balance between the duration of the machine and efficient use of diesel.

    7 kW / 0.70 = 10 kW

3. We derate the generator to take into account the installation conditions using generic values:

    Altitude: $\dfrac{800m - 100m}{100m} * 1\% = 7\%$

    Temperature: 30 °C – 25 °C * 0.2% = 1%

    Humidity: $\dfrac{40\% - 30\%}{10\%} * 1.5\% = 1.5\%$

    The total derating is: 10 kW (0.07+0.01+0.015) = 0.95 kW

    The **prime power** required is: 10 kW + 0.95 kW = **10.95 kW**

As it is unlikely for the manufacture of a 10.95 kW generator exactly, select the next size up.

Sizing for consumption with peak

There is a proposal to pump water using a 7.5 kW Grundfos pump powered by a generator. A particular brand of generator has been chosen which recommends derating the generator by 1% for every 100 m above 250 m, 2% for each 11 °C above 25 °C and 1.5% for every 10% relative humidity above 40%. The well is near Ouagadougou at 305 metres above sea level. The climate diagram for this region is as follows:

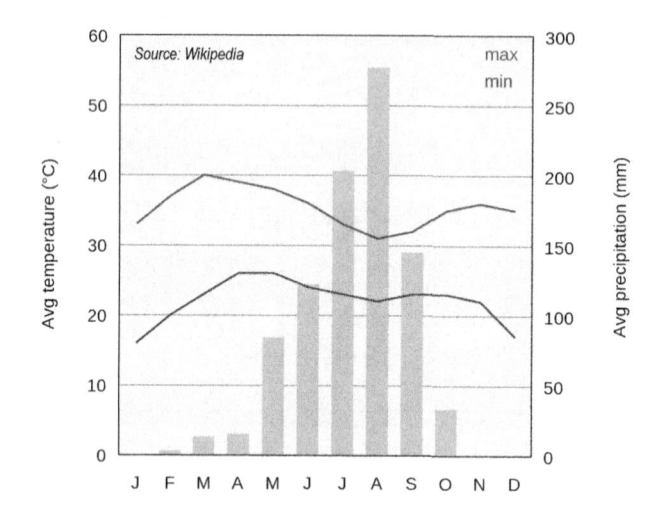

Relative humidity (%)												
Jan	Feb	Mar	Apr	May	Jun	Jul	Aug	Sep	Oct	Nov	Dec
19	19	20	28	40	49	62	67	60	44	30	23

1. The first thing is to determine the start-up peak. This data is given to you by the pump manufacturer as this depends on the type of construction that the pump has. From the Grundfos tables in Annex C, a factor of 2.0 is obtained.

Pump power requirement		Minimum genset power		Factor
HP	kW	kW	kVA	
4	3	8	10	2.7
5.5	4	10	12.5	2.5
7.5	5.5	12.5	15.6	2.3
10	7.5	15	18.8	2.0 ←
12.5	9.2	18.8	23.5	2.0

Take into account the start-up peak of the generator, which will be:

7.5 kW * 2.0 = 15 kW

Once the peak has passed it will operate at approximately: $\frac{7.5 kW}{15 kW} * 100 = 50\%$

50% is a good load for a generator. The load is an approximate because the pump consumes around 7.5 kW depending on the pumping conditions, not necessarily 7.5 kW exactly.

2. We derate the generator to take into account the installation conditions using the manufacturer's values.

Altitude: $\frac{305m - 250m}{100m} * 1\% = 0.55\%$

We are not going to derate for the temperature, as the start-up peak occurs when the machine has recently been warmed up, and after this the generator will have some excess capacity because it works at 43%.

The worst month for humidity is August: $\frac{67\% - 30\%}{10\%} * 1.5\% = 5.55\%$

The total derating is: 15 kW (0.0055+0.055) = 0.91 kW

The prime power required is: 15 kW + 0.91 kW = **15.91 kW**

If there are several motors to start-up, start the largest one first followed by the resting ones in such a way so that their peaks stay below the first peak. If the pump is started before other systems, e.g. lighting, **its consumption can be nested below the peak** thereby increasing the efficiency with which the generator works and avoiding the purchase of unnecessarily large generators.

52 Generators in development projects.

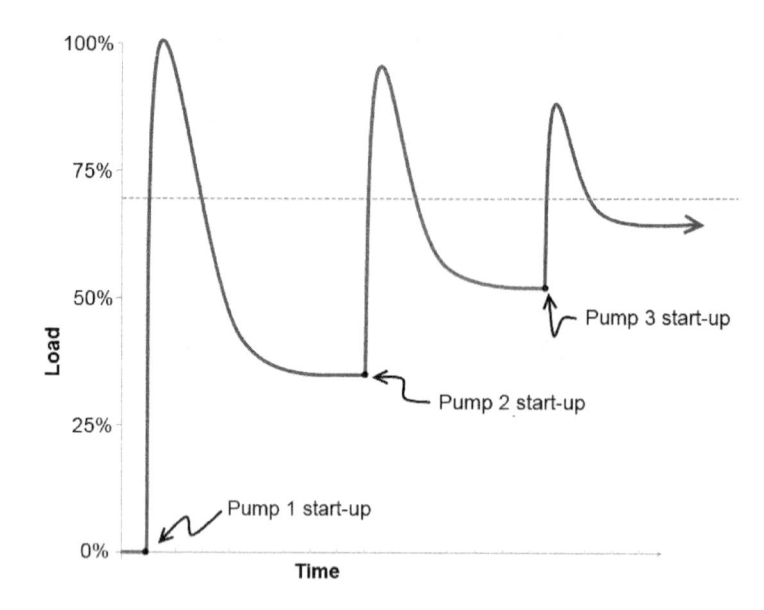

Fig. 2.9 Sequential start-up of motors without exceeding the first peak.

Not planning the start-up of a motor in relation to other consumptions activities is a waste of resources, as larger generators are needed that work innefficiently at lower loads.

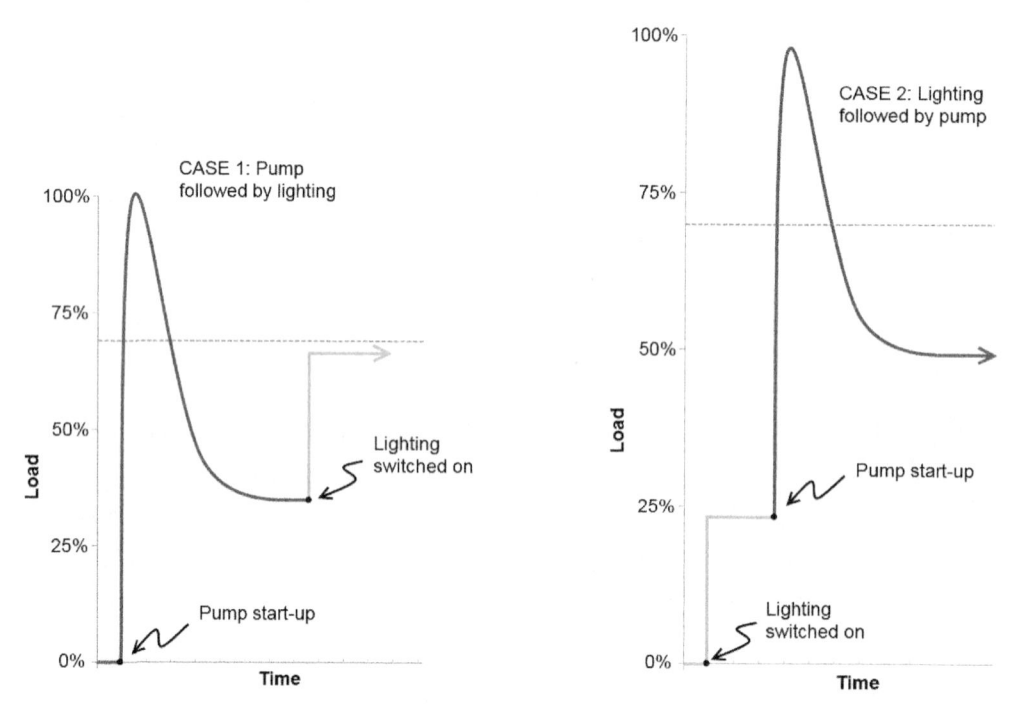

Fig. 2.10 Start-up below the peak vs. Start-up over the peak.

Sizing of backup for power cuts

A backup generator is to be installed in a hospital to help cope with the daily power cuts which often last several hours. The hospital is at sea level, the highest temperature is 32 °C and humidity is very low. The power consumption of the most important hospital equipment has been measured according to the time of day with the following results in kW:

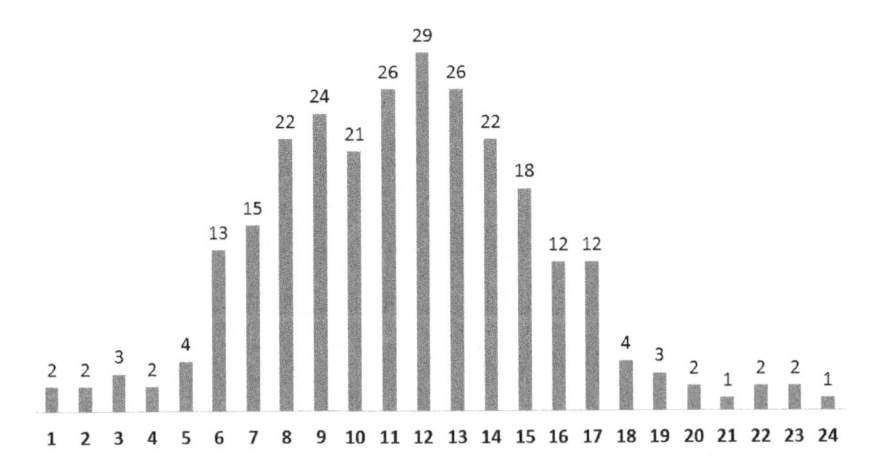

[Bar chart showing load in kW by hour (1-24): 2, 2, 3, 2, 4, 13, 15, 22, 24, 21, 26, 29, 26, 22, 18, 12, 12, 4, 3, 2, 1, 2, 2, 1]

1. To simplify the exercise, assume that the load distribution has already been optimized. For the distribution, it appears that there are two distinct operating times during the day, one of high consumption from 6 to 17 hours and another of lower consumption for the remainder of the day. To have a very large generator at low load is very inefficient, so having two generators, one bearing the largest burden during the day and another smaller one to cover the night, may be a solution.

2. The maximum load of the generator is 29 kW. We calculate the average load for the generator assuming it is 29 kW. For example, for 6:00:

$$\% = \frac{13\text{kW}}{29\text{kW}} * 100 = 44.8\% \approx 45\%$$

Time	6	7	8	9	10	11	12	13	14	15	16	17
Load kW	13	15	22	24	21	26	29	26	22	18	12	12
Load %	45	52	76	83	72	90	100	90	76	62	41	41

$$Al = \frac{45*1+52*1+76*2+83*1+72*1+90*2+100*1+62*1+41*2}{1+1+2+1+1+2+1+1+2} = 69\%$$

3. The derating for temperature is therefore: 32 °C – 25 °C * 0.2% = 1.4%

The prime power of the generator after the derating, since we don't use the standby rating as it will run more than 200 h per year, is:

$$29 \text{ kW} * 1.014 = 29.4 \text{ kW}$$

4. For the second generator the same process is followed, keeping in mind that the loads under 30% are calculated as 30%:

Time	18	19	20	21	22	23	24	1	2	3	4	5
Load kW	4	3	2	1	2	2	1	2	2	3	2	4
Load %	100	75	50	25	50	50	25	50	50	75	50	100

$$AL = \frac{100*2 + 75*2 + 50*6 + 30*2}{2+2+6+2} = 59\%$$

Therefore the second generator is 4 kW prime. It is not derated because the value would be ridiculously low, at barely 50 W.

5. Check that it will not have motor start-up peaks that the generator can not cope with. For example, if part of the consumption is a 4 kW pump with a 2.5 startup peak, the pump can not be started up when this consumption is greater than:

29 kW - 4 kW* 2.5 = 19 kW

In this case, you could start up the pump from 6:00 AM until 8:00 AM and from 15:00 PM a 18:00 PM, or consider a larger generator.

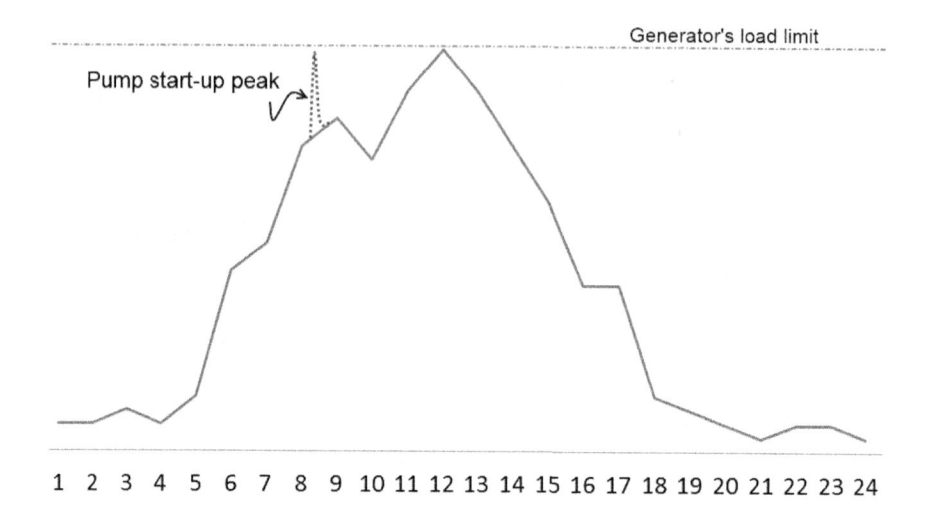

Sizing of a supply generator

A generator is being sized for the only health centre on one small island in Guinea Bissau which has a population of 7,000 people. The island is at sea level, humidity is very low and the generator is specified for 40°C. The devices that consume electricity on this particular health center have been measured for their real consumption giving the following results:

Water Pump: 3 kW
Autoclave: 2.3 kW
Night lighting: 44 W
1 kW for the office
Fridge: 135 W
X-ray machine: 3.5 kW
Washing machine: 1170 W

This case is very similar to the previous example except that, instead of measuring the power consumption, we must determine and structure consumption by looking at when each device is used.

1. A meeting is arranged with the people involved to agree on a consolidated schedule of use for a generator at a medical centre, that has a good balance between keeping operating costs to a minimum and facilitating the work of the centre. The usual tendency is to want electricity instantly and on demand, so it is very important that people understand the costs associated with how and when this consumption takes place.

    In the meeting the following points were agreed:

    a. The water pump must work three hours every day, but the timetable is completely open as to when this might occur.
    b. The autoclave will be used at the end of the workday when the instruments are washed and sterilized for the next day. This sterilization process requires two cycles of 60 minutes each.
    c. The night lighting and the fridge are essential but, so that they work automatically without supervision, a bank of batteries will be used. Office hours are from 8:00 to 12:00 and 14:00 to 17:00.
    d. The radiographs will be made in a flexible two hour period.
    e. Three 1 hour cycles of the washing machine are needed daily; they would be better to occur in the morning to allow for air drying of clothes.

56 Generators in development projects.

You have calculated that you need to store 2.7 kW and that the charger consumes 445 W during 8 hours to charge the batteries.

2. The above points are consolidated on a usage schedule in the most practical way possible. One possible solution would be the following:

Load	kW	1	2	3	4	5	6	7	8	9	10	11	12	13	14	15	16	17	18	19	20	21	22	23	24
Water pumping	3														■	■	■								
Autoclave	2.3																	■	■						
Lighting	0.044	■					■	■										■	■	■	■	■	■	■	■
Office	1								■	■	■	■	■	■											
Refrigerator	0.135	■	■	■	■	■	■	■	■	■	■	■	■	■	■	■	■	■	■	■	■	■	■	■	■
Washing machine	1.170								■	■															
X-ray machine	3.5										■	■													
Inverter charger	0.445																	■	■						
TOTAL HOURLY CONSUMPTION		0.18	0.18	0.18	0.18	0.18	0.18	0.18	2.79	2.79	6.29	5.12	0.18	0.18	4.62	4.62	4.62	2.92	2.92	0.18	0.18	0.18	0.18	0.18	0.18

The load pattern of the generator and the battery bank would therefore be the following:

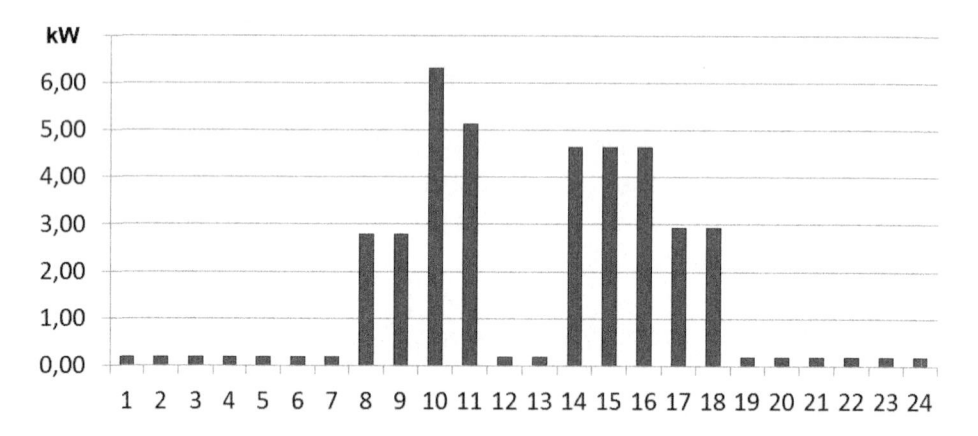

3. We must now check the start-up peaks. Note that the water pump has been put at the beginning of the second shift of the generator to turn it on on its own, followed by the the rest of the loads. The start-up peak of a 3 kW pump is 2.7 according to the table at Annex C. The minimum power required for the generator to start the pump is:

3 kW * 2.7 = 8.1 kW

© Santiago Arnalich   Arnalich. Water and habitat   www.arnalich.com

4. At the time of largest load, 10 AM, 6.25 kW is consumed. As this load is less than the pump peak, this is what determines the generator power.

5. It is not necessary to apply a derating for working conditions. The generator that we need will have 8.1 kW of prime power.

6. There are 9 KW generators available locally. The average load of one of these generators, if installed, would be:

$$Al = \frac{30.56*2 + 69.44*1 + 56.44*1 + 50.89*3 + 32*1 + 32.49*1}{2+1+1+3+1+1} = 44.91\%$$

7. The last step would be to determine the operating cost in the same way as was set out in the first chapter of this book and to evaluate if these costs are manageable.

Often, expenses that are at first considered necessary are not affordable and compromises must be made. In such cases, **the generator is sized based on the annual budget.**

3

Selection and purchase

Due to the peculiarities of development contexts, bear in mind that:

- It is better to order **simple models and options** which are most common in your area and that can be repaired with the skills, tools and spare parts available locally. Avoid highly sophisticated motorizations with delicate control systems and sensors, as well as generators with parts that can be deprogrammed.

- A purchase order for **non-standard machines or accessories can delay the sending and receiving of spare parts by many months**. If the need for the generator is truly necessary, it would be difficult to wait more than a few days.

Understanding purchase options and specifications

Once you've determined the power required for your project, it is important to determine the available options before deciding on the best generator for your particular needs. This chapter reviews the main specifications and options available to help you make an informed decision.

Voltage, frequency and number of phases
Options : Three-phase or single phase

The voltage and frequency are determined by your local area and the products you are able to buy. What is left for alternatives (only in low power machines) is whether to purchase a single or three phase generator.

You can think of the phases like the cylinders of an engine. In a single phase system, for each oscillation electricity vanishes at some point where it runs out of power. In three-phase systems, although one phase will void, the rest are still strong. This means that single phase machines have to be adapted to be without power for each cycle and must have more inertia, since they have much heavier and more expensive motors and installations that use more copper.

Single-phase generators are simpler and avoid the complication of having to manage three phases. They can be connected directly without modification to the network of a building. If the load involves many motors, three-phase is a better choice.

Type of response to the load
Options: G1, G2, G3 and G4

The application of loads slows the generator and varies both the frequency and the voltage until it adapts, as you saw in an earlier section. Some generators have a quicker and more accurate response to the loads than others, depending on their planned use (see table below). Most applications require a G2 type. For example, to connect the computer that you connect at home, you do not need a G4 type, a G2 would do, since you connect your computer at home. When in doubt, it is best to ask the manufacturer.

Sometimes the manufacturer gives a recommendation. For example, X-ray machines usually require that the voltage drop is not greater than 10%. In these cases consult the generator specifications to make sure there are no problems.

Type	Description	Examples
G1	Applications without major requirements	Lighting, heating with resistors
G2	Level of requirement similar to that of a connection to the utility	Lighting, pumps, appliance with motors, etc. Things you would connect at home.
G3	Applications with a strict demand for voltage, frequency and wave form	Specialized telecommunications equipment and those with thyristors
G4	Applications with exceptionally severe demands	Data processing equipment and computers (industrial scale)

Fig. 3.1 Types of response according to ISO 8528-1.

Type of insulation and temperature increase in the alternator

Options: Combinations of A,B,F and H

The specifications are represented by two letters separated by a forward slash (such as F/B, H/F or H/H). The first letter represents the insulation type and the second letter represents the temperature increase. There is greater insulation or temperature increase as the letters progress in the alphabet.

For the same generator, alternators with a lesser temperature increase allow for a better motor start-up, lower voltage drops and greater ability to accept non-linear or unbalanced loads. Also, the higher the type of insulation over that of the temperature, the longer the duration of the insulation. In other words, H/B is better than H/H.

Excitation Type

Options: Permanent Magnet (PMG) or self-excited

Excitation type refers to how the magnetic field needed for the generation of electricity is created. In the case of a self-excited generator it is the alternator itself that produces the magnetic field while the permanent magnet type has magnets that maintain the stability of the magnetic field.

The fundamental difference is that in the self-excited models, if the generator changes the rotation speed (for example as it slows down during the start-up of a big motor), this decrease in speed weakens the magnetic field and reduces the voltage. Furthermore, when these generators are turned off they depend on a weak residual magnetism so that they can not start-up with a load. When they lose the residual magnetism it must be reset with external currents.

On the other hand, a self-excited generator is self-protected against excessive loads because these loads collapse the magnetic field and end the induction. A permanent magnet generator allows current surges of two or three times its nominal load for up to ten seconds. This is very useful for starting up motors and non-linear sources, but needs external protections so as to not self-destruct.

Motor Revolutions

Options: 1,500 or 3,000 rpm (1,800 or 3,600 rpm in generators of 60 Hz)

If the speed is increased from 1,500 to 3,000 rpm (or from 1,800 to 3,600 in 60 Hz) machines of the same power but half the size are achieved. Generators with this configuration are cheaper and smaller, but bring some disadvantages: more noise with an unpleasant tone, shorter lifespan, more frequent maintenance and poorer ability to start up motors.

Air intake type

Options: Atmospheric or turbocharged

The motors with turbo are comparatively smaller and are more limited for accepting very large loads.

Speed regulator type

Options: Mechanical or electronic

Regulators or speed controllers control the input of fuel to the motor to maintain it at the correct revolutions with different loads. The mechanical regulators mechanically detect the revolutions, usually keeping them in ± 3-5% between unladen to full load. They are used for applications where the frequency drop is not a problem or where there is not the capacity for more sophisticated maintenance.

The electronic speed regulators are much faster and increasingly dominant because of the rules to control emissions. It is probably not possible to choose mechanical speed regulators in some generators.

Start-up Systems

Options: Electric or pneumatic; Simple, remote or automatic.

Manual cord start-up systems are installed in very low power generators and compressed air systems are often more convenient for more powerful generators than those considered in this book. Start up by battery is better known and more common. Due to the draconian rules of some applications, some generators specify a 10 second start, a cold load and other applications that can greatly complicate and increase the expense of the installation.

If the start-up is simple, consider if a key is necessary or whether a switch is enough. Remote starting or the automatic power failure starting function require accessories that must be ordered.

Generators with cabin
Options: IP Grade protection, acoustic protection, material.

The cabin is primarily used to protect the generator from the elements in a compact manner and/or to reduce noise, but does not retain the heat for long. In conditions where corrosion is expected the cabins need to be made of aluminum. A generator with cabin a should not be installed within an enclosed building to avoid excessive restrictions of air flows.

Fig. 3.2 Pramac GBW22 generator with cabin.

Fig. 3.3 Pramac GBW22 generator with cabin opened for maintenance.

Noise measurements will typically be given in decibels at a distance of 7 m. In Chapter 4, noise is described in more detail.

The cabin protection is measured with the IP standard, e.g. IP32. The first digit is protection against the entry of solids and the second against water.

IP	Entry of solids and objects	Entry of liquids
0	No protection	No protection
1	Foreign bodies with a diameter >50mm	Vertical drops
2	Foreign bodies with a diameter >12mm	Dripping water when tilted up to 15°
3	Foreign bodies with a diameter >2.5mm	Water spray (up to 60 ° from vertical)
4	Foreign bodies with a diameter >1mm	Water spray
5	Contact and dust deposits on the inside	Water jets (from all directions)
6	Contact and dust penetration	Accidental water injection
7	N/A	Temporary immersion
8	N/A	Indefinite immersion
9	N/A	Pressure injection of water (80-100 bar)

Fig. 3.4 IP code breakdown chart.

Control Panel

Options: Analog, digital, other; Integrated or separate housing

The analog panel is the oldest type of panel and has needle indicators, like the one shown in the introduction. There is usually a similar digital version which has some alarm or extra indicator added. Other more advanced digital panels (with a microprocessor or total control) may only bring problems in contexts where they will have a hard life. Whichever panel you choose, it must always have a **red emergency stop button**!

Fig. 3.5 Integrated analog panel, with emergency stop button.

An analog or digital panel which is as simple as possible, is preferable so as to avoid having to take the generator to a "psychologist" to solve the flood of ghost sensor failures and to be able to organize local repairs with pieces of scrap, as seen on the next page.

Fig. 3.6 Integrated analog panel with emergency stop button on another damaged panel.

Tropicalisation?

Sometimes tropicalised generators are offered, without much information or consensus regarding what they involve or when they are needed. Sometimes these generators have had a treatment against fungi, in other cases they include a anti-humidity treatment which may also be required in non-tropical climates. Ask the supplier exactly what features tropicalisation offers.

Ordering the generator and accessories

When you are clear about the power requirements and the basic specifications of the generator, it is time to talk to the suppliers to see what they have available that might meet your particular needs.

Do not leave this transaction solely to your colleagues from logistics. You've seen that generators are not always the same despite having the same power. The purchase price is not the only criterion and can lead to false economies (bigger is not better, in fact it is the opposite). It would not be the first time a generator model has been purchased simply because it was on sale as though it were a sack of potatoes!

During these first discussions with the supplier, it is important to be clear on:

- The power required.
- The type of use: prime or standby.
- The voltage, frequency and number of phases.
- If the generator will have a cabin or if it will be open.
- The weather and temperatures at which the generator will operate.
- The type of control panel.
- The level of tolerable noise.
- The hours that the fuel tank should last on a daily basis.
- The type of equipment that will be connected and their tolerance to frequency and voltage variations.
- The start-up load of motors.
- The maximum load that will be connected for each step of the sequence.

The accessories

In addition to the generator you will need some of these accessories:

- Independent battery charger. Starting the generator with cables is dangerous because sparks could ignite accumulations of hydrogen.
- In-skid reservoir tank and/or external tanks.
- Antivibration spring mounts for the noise.
- Coolant heaters in cold climates or when the generator must comply with a ten second start standard.

- Oil cooling heaters in very cold climates.
- Anti-condensation heaters in cold or extremely humid climates.
- Exhaust pipes and their accessories.
- Hot air discharge duct (very important!).
- Transfer switch.
- Grounding material.
- Energy meter or current clamp to understand the consumption of different appliances and decide how to use them.
- Fire extinguishers (either an ABC or BC type).

Remember also to ask for the user manual and the spare parts catalogue!

Essential spare parts

Check with the supplier for the essential spare parts of the generator, and in particular those that meet all of these conditions:

- They are not part of regular maintenance.
- They are more likely to fail.
- The generator is rendered useless when they fail.
- They are not expensive. It makes no sense to store an entire replacement alternator, but it does make sense to keep an extra fuse or a diode on hand.

Ordering spare parts

Imagine that you need to order an hourmeter for a Pramac S12000 generator. The first thing to do would be to identify it in the spare parts book:

S12000 GENERATOR PARTS

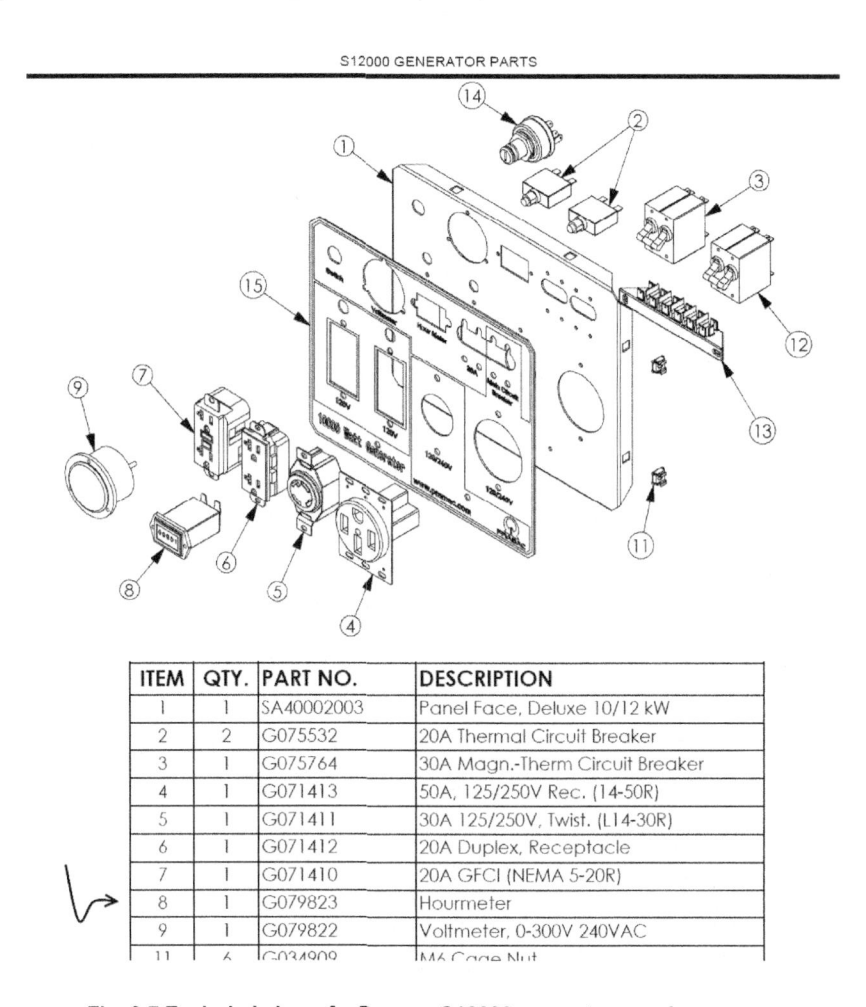

ITEM	QTY.	PART NO.	DESCRIPTION
1	1	SA40002003	Panel Face, Deluxe 10/12 kW
2	2	G075532	20A Thermal Circuit Breaker
3	1	G075764	30A Magn.-Therm Circuit Breaker
4	1	G071413	50A, 125/250V Rec. (14-50R)
5	1	G071411	30A 125/250V, Twist. (L14-30R)
6	1	G071412	20A Duplex, Receptacle
7	1	G071410	20A GFCI (NEMA 5-20R)
8	1	G079823	Hourmeter
9	1	G079822	Voltmeter, 0-300V 240VAC
11	6	G034909	M6 Cage Nut

Fig. 3.7 Exploded view of a Pramac S12000 generator panel.

Then you would make the order specifying the quantity you need, the part number and the description. You should also include the model and serial number of your generator (if this request is accompanied by a photo of the nameplate, even better).

For example:

1 Hourmeter (Part No. G079823) for a Pramac S12000 generator with serial number 802893-323-B.

4

Installation

Check the local standards relating to generators, especially backup generators, regarding fire protection, fuel storage and others standards that may be applicable in your region.

The installation of a generator requires professional electricians that are familiar with the work and who know the local rules that must be respected.

Make sure that the generator, protections and appliances all work with the same frequency.

Pay special attention to fulfilling all the requirements presented in this chapter no matter how trivial they may seem to you and no matter how thin you are spread over the other infinite emergencies and tasks that plague you. In development projects these requirements are rarely honoured, and it is one of the main reasons that generators cause some impossible expenses, have short lives and break down all too frequently. Giving special consideration to these matters at the start of your project will save you endless headaches and costs later on.

Selecting the site

The generator should be in a place that does not flood, is not susceptible to damage from disasters, vandalism, theft or armed conflict and that accommodates two principles that are often contradictory: to be as close as possible to the loads to avoid power losses while being as far as possible from people because of the noise and toxic gases.

Some other points to keep in mind:

- Determine the access you have for filling the fuel tank, for putting in and removing the generator and for carrying out future replacements and overhauls in situ.

- Find places that will not be a hindrance to other uses or expansions of existing infrastructure.

- Find a site with well-defined land ownership, where all the installations are made in places that will not compromise site access or be susceptible to a change due to people's opinions, institutions or conflicts between communities.

- In the case of backup generators, look for places and configurations where a disaster will not affect the main and backup installations at the same time.

- Makes sure the grounding is sufficiently distanced from other groundings to avoid interferences (at least 20 m).

- Avoid locations that pose a fire hazard or would endanger people if they ignite (for example, on the ground floor next to the only way out of a building).

- Find locations that allow for consensus among people.

- Avoid proximity to sources of water that may become contaminated with oil and fuel spills.

Once you have identified some options, make a sketch that summarizes the installation, the accesses, the inlets and outlets of air and gases, the noise and other things that you will see later in this chapter. It is better to find the errors through a sketch on paper than in the concrete! Once you have done this, you can go on to seek the approval of all the parties involved.

Transport of generators

Whether at the beginning of the installation or when transporting the generator for a change of site or for maintenance, remember these two points that often damage generators:

1. Do not transport generators in open vehicles without **fully protecting them with a tarpaulin** to prevent damage from rain, dust and impact of insects or sand.

2. When lifting the generator with a hoist, **use the anchor points on the base**; the generator must never be attached by the motor or the alternator anchors as they would lose the alignment between them. Check that none of the slings touch any part of the generator when in tension so as to avoid crushing it. This can be achieved with dividers that act like puppet strings. Check that the hoist is available beforehand to avoid improvising a bad solution halfway through the job!

In the photo the anchor points are being used, but notice how one of the slings is crushing the exhaust:

Fig. 4.1 Hoisting the generator in Adado, Somalia. The sling is compressing the exhaust.

The generator house

After selecting the site, you must define the type of generator house or the room that will house the generator. When the generator has its own cabin and the climate is suitable, an elaborate structure is not necessary, just a shelter to protect it from the sun and rain, and make the work of the operators more bearable. The cabin will already have taken into account the requirements of ventilation, soundproofing, and so on.

Fig. 4.2 Covering over a generator with cabin in Nairobi, Kenya.

You should not install generators with a cabin, especially the acoustics type, in enclosed spaces due to the excessive restriction of airflow and incompletely sealed exhausts that can cause poisoning.

Likewise, **you should not install open generators in open sheds**, since they must be protected from the elements, and especially, the wind in arid and dusty climates. The deposited dust compromises the cooling and the sand blown hard by the wind erodes the insulation. The generator below, for example, has been badly housed.

Fig. 4.3 Inappropriate shelter for an open generator, Wargalo, Somalia.

Sometimes you must also plan a space for storage or for the operators. The generator house should not be used to store other things because of the increased risk of fire, accidents and corrosion (for example, near chlorine).

Fig. 4.4 House with storeroom in Mudug, Somalia.

The building in the photo above could be improved if the door, which serves as the air inlet, had louvers and if the roof were made of a material that radiates less heat to the generator than the corrugated iron. If the corrugated iron roofing is already there, make holes for ventilation immediately below the sheeting on its highest side.

Dimensions

The overall dimensions of the house and the door should be large enough so that you can take the generator in and out and perform all the maintenance operations. If it is a generator with a cabin, make sure that the doors can be opened. There needs to be at least one metre of space around all sides of the generator. Plan a gentle ramp for easy entry and exit of materials. Annex E summarizes the tentative dimensions of these various elements.

Exhaust planning

The exhaust pipe should be as simple, short and straight as possible to evacuate the gases easily. If the generator has elbows, they should be elbows with a curvature radius of at least six times the diameter. is the exhaust pipe must be installed as high as the house as possible to help disperse the smoke, and should discharge away from windows, doors, occupied buildings, air intakes or flammable material. Where possible, the exhaust pipe should also be aligned in favour of the prevailing wind direction. The installation must take into account the efficient transit of discharges and gases as explained below.

To prevent corrosion by water condensation, the exhaust pipe should be installed with a slight inclination towards the outside, with the silencer as close to the motor as possible and with accessories for water evacuation. The exhaust must accommodate the thermal expansion, and be hung so that its weight does not rest directly on the motor outlet. Remember to protect the exhaust with a grille to prevent birds from nesting if the output is horizontal. The most powerful silencers are very bulky and usually have special space requirements.

Air Inlet and outlet.

The air inlets and outlets should be large enough so as to not impede the flow of air. Normally the inlet is 1.5 to 2 times larger and its shape is more flexible. The outlet is adapted to the shape of the radiator, and as it is a significant source of noise should be oriented so as not to cause a nuisance. In planning, keep in mind that the outlet should have a hot air discharge duct from the radiator to the wall as explained later in the section "Flows of air and gases".

Installing **louvers is very important** to avoid premature deterioration of the generator due to dirt, wind, sand and rain. Although in a development context it is common to install bars instead of a door as you can see in the picture, because this is easier or cheaper, however this should be strictly avoided. Doors should act as a physical seal covering the door opening and not only prevent access of people.

Fig. 4.5 Bars in place of door and louvered inlet, a frequent mistake in hot climates.

Protection against the risk of war

In areas of conflict it may also be necessary to protect generators with sandbags within the enclosures of hospitals, offices, living areas, and other places. The book *Staying Alive* from the ICRC provides good details on how to do this effectively.

Foundation

The generator should be supported on a structure that is sufficiently solid to maintain the alignment between the motor shaft and the alternator.

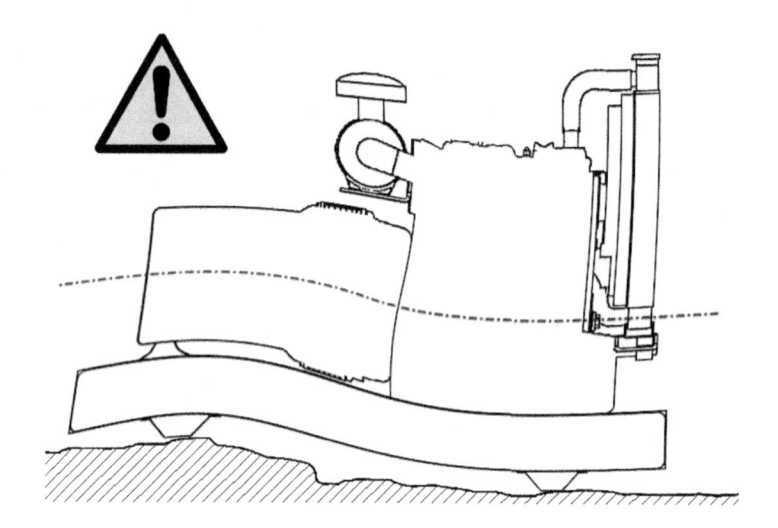

Fig. 4.6 Badly supported generators distort the alignment of the shafts.

The fastest and most convenient way to provide a stable foundation for a generator is to place it on a concrete slab. The width and length of the slab should exceed the generator by at least 20 cm on all sides, and weigh at least two times the weight of the generator in operation. To achieve, this you can calculate the thickness of the slab according to the formula:

$$T = \frac{2*k}{w*l*2400}$$

Where,   k = weight of generator in operation
w = width in metres
l = length en metres
2,400 kg/m^3 corresponds to the density of the concrete 1:2:3.

It is very important to bear in mind that:

- The foundation must be isolated from the rest of the building's foundations to prevent damage to other structures through vibration.

- The generator must not be placed on top of any combustible material.

- If you are using an existing structure, make sure it is able to sustain the generator, the accessories and the reservoir tank.

- If the foundation is going directly on the floor, ensure that the floor has the capacity to bear the load.

- The foundation can not be directly supported on rock, cement or iron because these materials transmit vibrations over great distances. To avoid this, the concrete slab must be placed on a layer of sand of at least 20 cm.

- The foundation must be isolated from other structures with machines because, when these are stopped, they do not have oil pressure to lubricate the components and the vibrations wreak havoc.

- The platform should be raised by at least 15 cm for easy maintenance (drainage and access), to prevent flood damage and to prevent water condensation collecting below that could get sucked in by the motor or alternator.

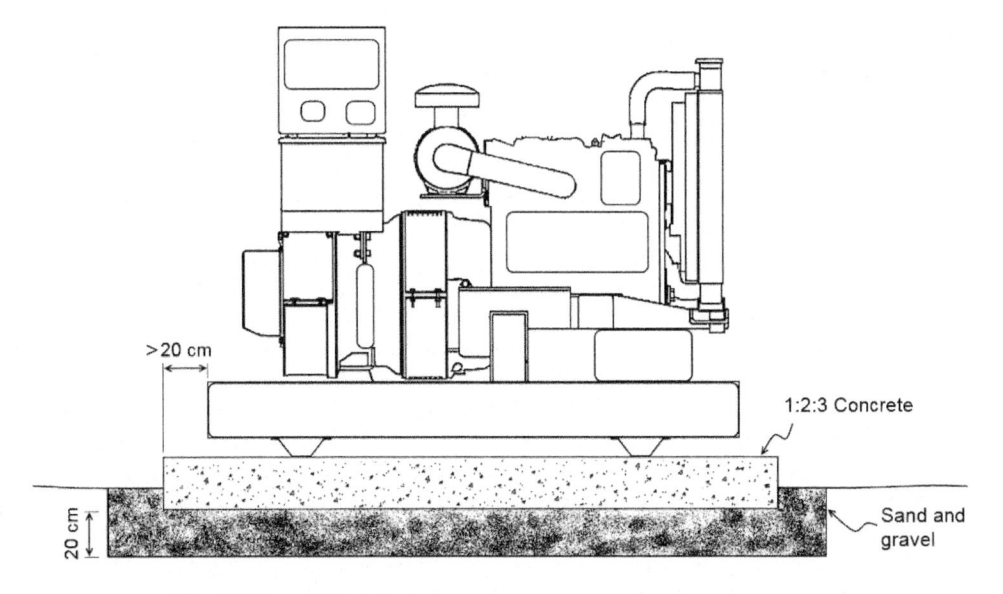

Fig. 4.7 Foundation diagram.

Flow of air and gases

It is important that a generator takes in cold air and evacuates the hot air in opposite vents. Remember the derating of the generator due to temperature mentioned earlier in this book? If the gnerator house overheats the generator loses power and consumes more diesel.

To not organize the air inlets and outlets is a significant waste and quite absurd given that the preventive measures are very easy and cheap:

- Try to install the **inlets and outlets on opposite sides**, with the air entry as low as possible and the air exit at the the height of the radiator and in favour of the prevailing wind.

- Install a **hot air discharge duct** with bellows coming out of the radiator to avoid the hot air staying in the house. The duct should be at least 1.25 times the area of the radiator. Installing this ductwork in development projects is as rare as it is vital!

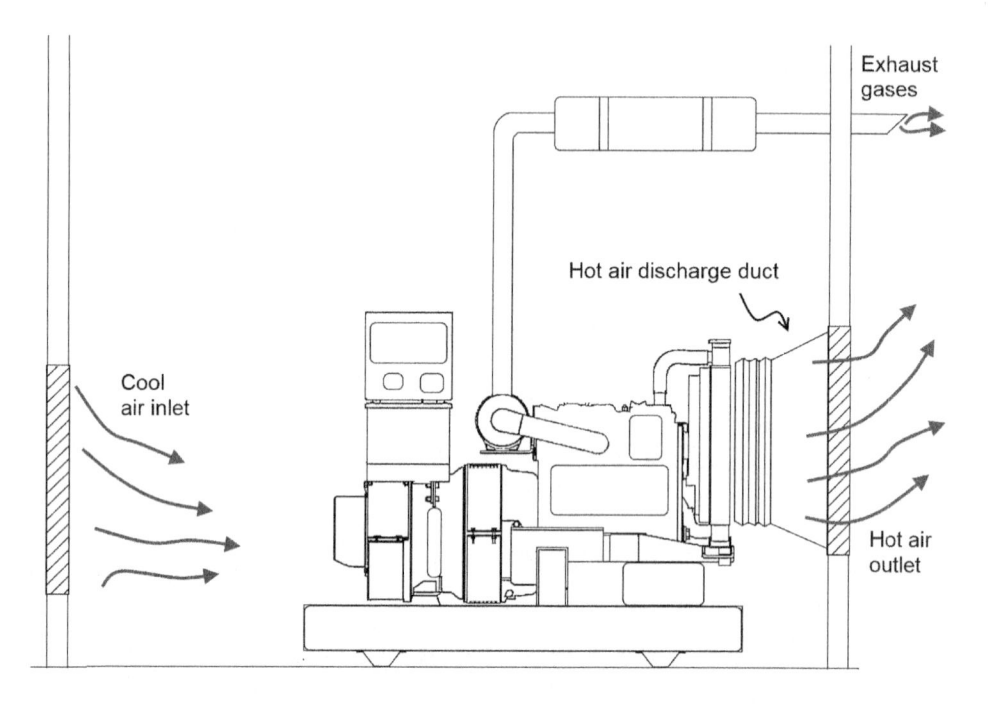

Fig. 4.8 Correct flow of air and exhaust. Outlet with hot air discharge duct.

- Make sure that the arrangement of the entrance and exit **forces air circulation through the body of the generator**:

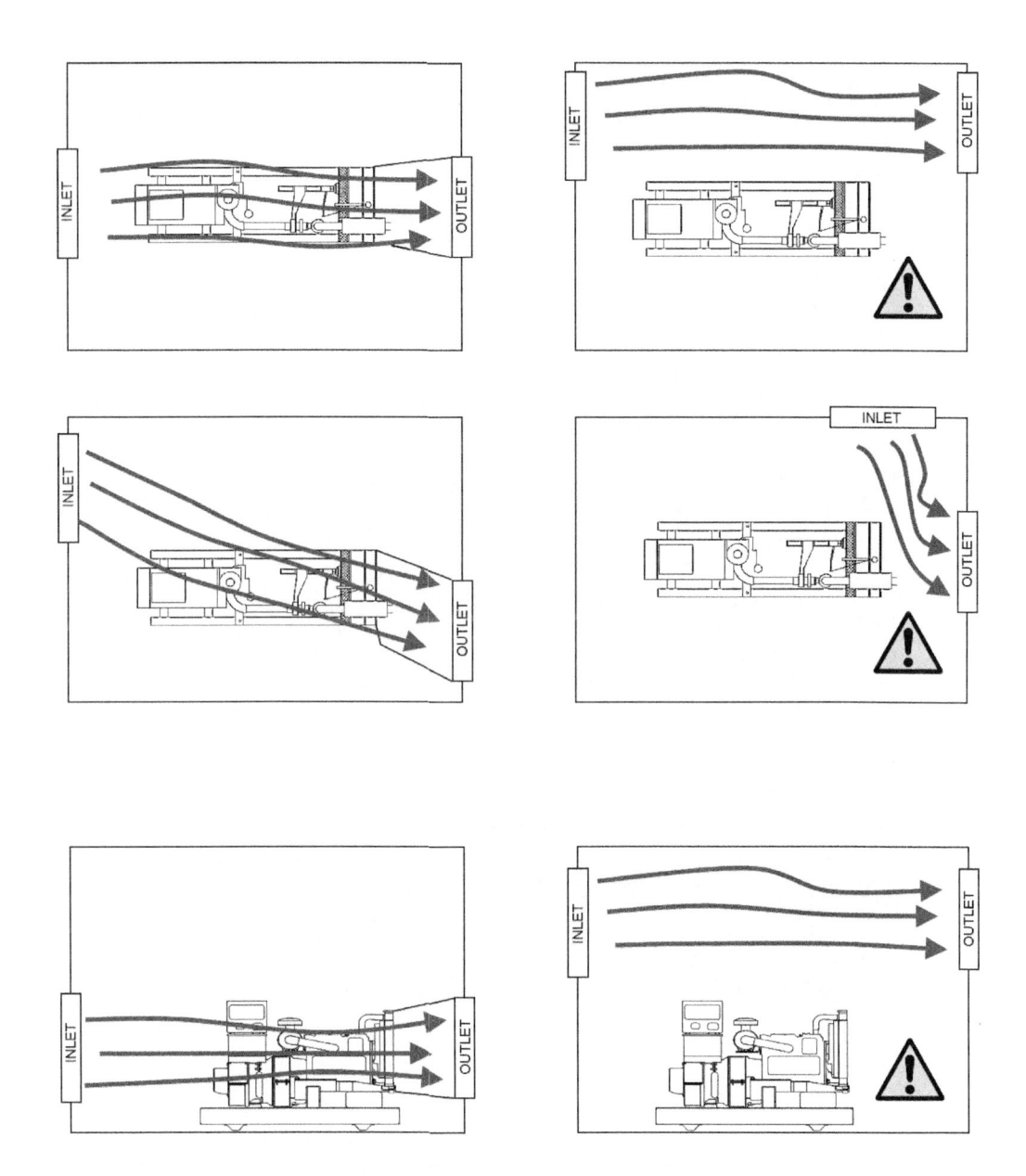

Fig. 4.8 The provisions on the right do not guarantee adequate cooling.

Remember the derating for altitude due to lack of air mentioned earlier in this book? The outlet for the exhaust gases should also be far from the inlet, not only because of the temperature but also to prevent smoke particles from re-entering and saturating the air filter prematurely. With the air filter clogged, the consumption soars and power is reduced.

82   Generators in development projects.

Fig. 4.9 Detail of the hot air discharge duct in a generator with cabin.

Ideally the air flow in the house should follow the prevailing wind direction. In windy locations or those with variable wind, where you can not properly orient the generator according to the wind or simply do not know (because it is simple and cheap to build), you should raise a **barrier against the wind in front of the outlet** similar to that described below for noise, at a distance of at least one times (ideally three times) the height of the radiator.

Fuel tanks

Often generators come with the fuel tank built into the base in what is called a day tank. The tank should be large enough to supply the generator during the period of work given that the generator should not be refueled whilst hot or while running (and it is slow to cool down!), and that **you have to wait for the particles and the water to settle again after refueling**. This is the most compact arrangement for the tank and the one that requires the least work.

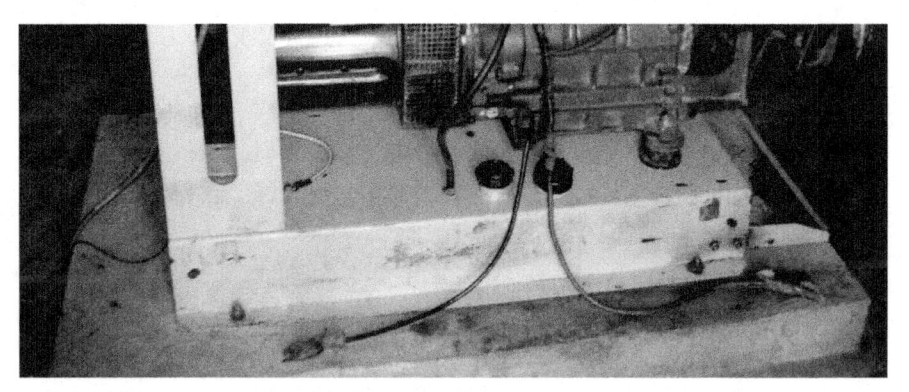

Fig. 4.10 Detail of a day tank built into the base.

The other two alternatives are an independent tank in combination with a day tank or an independent tank on its own. When this independent tank exists, it goes outside the house for easy filling and maintenance.

General considerations

- **Use black steel pipes and fittings.** None of the parts of the fuel system can be galvanized, brass, copper or zinc, all of which catalyze decomposition of the fuel.

- The inlet must be **50 mm above the bottom** of the fuel tank to avoid sucking in sediment, dirt and water.

- The fuel tank must have a **bottom drain valve** to remove water and sediment.

- Fuel tanks should not be completely filled. Usually **about 6% is needed for thermal expansion**.

- The double-walled tanks serve to contain leakage in the event that the first wall is punctured.

- The place where the tank and pipes are **should not be too warm, too cold, or have sudden temperature changes**. Diesel expands with heat and from 71 ºC causes a significant loss of engine power. 71 ºC is not hard to reach for a dark pipe exposed to the sun. If it's too cold, gelification of the diesel occurs and with too many changes of temperature, water condenses within the tank.

- Filling a fuel tank using a water separating filter prevents damage to the injectors and the injection pump.

In the case of an independent tank:

- The pipe run should be as short as possible, generally less than 6m.

- The pipe connecting to an independent tank must be equal to or greater in diameter than the pipe from the day tank, generally greater than ½" for the powers considered in this book.

- The fuel pipe must be protected from vibration, and should be far away from electrical cables, exhausts or hot parts.

- The fuel outlet will ideally be at a height between that of the entry of the injection pump and that of the injectors, so as to receive fuel but so that it does not overflow into the return lines from the engine.

- The path must be sloped continuously from the tank to the generator to feed by gravity and care must be taken to avoid areas that can trap bubbles of air.

- The tank must be vented to relieve excess pressure and to allow it to be emptied without creating a vacuum.

Noise reduction

Diesel generators are noisy, very noisy. While you may not need any attenuation (dampening) to control this noise in some situations, in others it is very important. In the majority of these cases, an attenuation of 35-40 decibels (dB) is sufficient. In addition to the noise level, the tone also matters. The 3,000 rpm generators, as well as being much noisier, emit a louder, more piercing noise which is much more annoying.

The decibel (dB) scale

The decibel measurement follows a logarithmic scale so that the double of 40 dB is not 80 dB but 43 dB. The noise level is doubled for every increase of 3-5 dB.

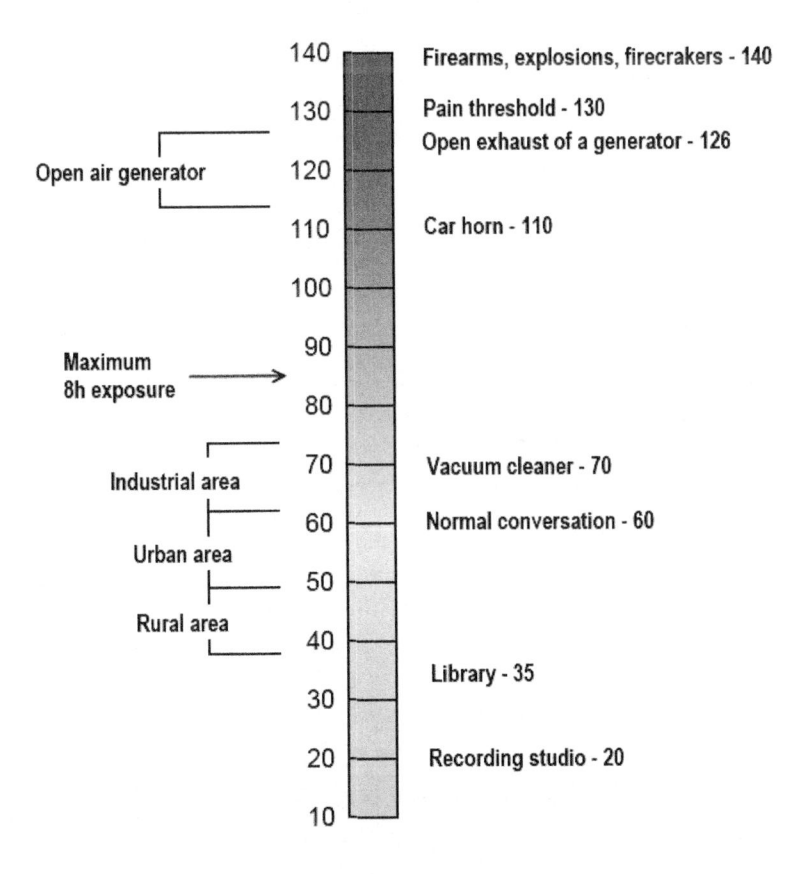

Fig. 4.11 Decibel scale with levels of common noises.

Sources of noise and attenuation strategies

Regardless of the source, the cheapest and easiest way to remove noise is to move it further away. Every time the distance is doubled, an attenuation of about 6 dB occurs. The interesting thing about this strategy is that most of the attenuation occurs in the first few metres of distance, so a little further can go a long way in the attenuation.

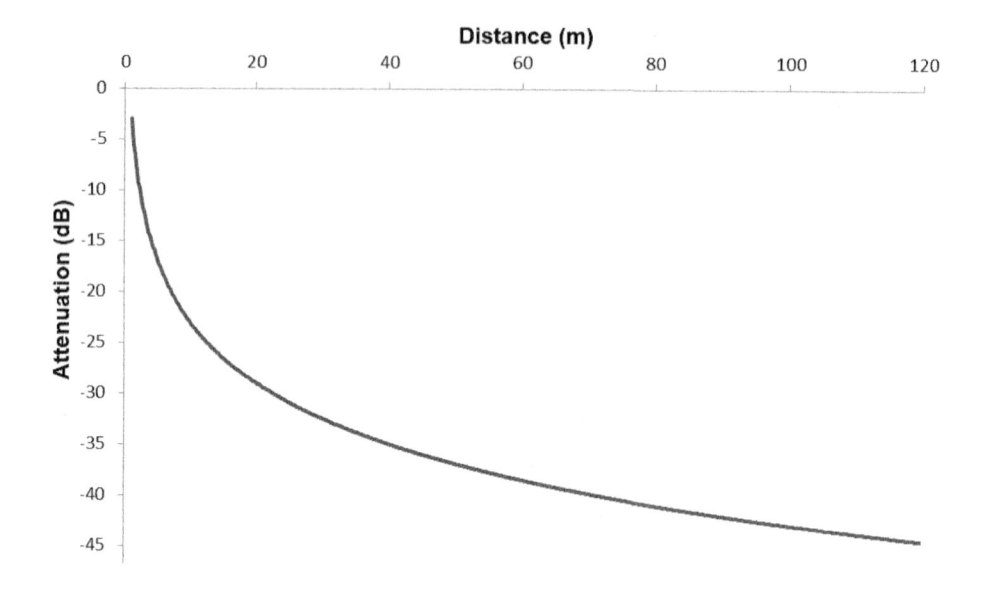

Fig. 4.12 Noise attenuation vs. distance.

A. **Exhaust noise.** It is one of the most pronounced noises, but is easily resolved with the installation of silencers. Silencers are typically manufactured with four levels of attenuation:

- **Industrial:** reduction of 12 to 18 dB.
- **Residential:** 18 to 25 dB.
- **Critical:** 25 to 35 dB.
- **Super-critical:** 35 to 42 dB

The most effective silencers can be very large, so you should first make sure that they will fit in the generator house. Note that moving a generator 40m away produces a similar reduction to a super-critical muffler.

B. **Mechanical noise.** This comes from the movement of the parts of the generator. Its attenuation is a question of mass between the source and the listener, so it is common to use concrete blocks filled with compacted sand or mortar in the construction of generator houses or walls for sound barriers.

Any grooves, cracks and access doors that seal poorly are escape routes for the noise and so it is important to correct them. There are soundproofing inlet and outlet louvers which can assist with this.. Although it is possible to line the inside of the house with sound-insulating material this is usually quite expensive. Another solution is to purchase attenuated cabins, which are expensive but can avoid having to build a solid house.

Fig. 4.13. Maintenance of a generator in an attenuated cabin. The soundproofing material is visible on the doors.

C. **Refrigeration noises** are those made by the radiator and fan. Although you can alleviate these by using oversized radiators or moving the radiator away, the most practical method in a development context is probably to simply construct a sound barrier of filled blocks. Make sure that the distance is at least 1.5 metres from the outlet so as to not compromise the cooling process.

D. **Noise vibration.** These are the most challenging noises to deal with, and are generally transmitted through the structures over long distances. Noise caused by vibrations are dealt with by insulating the generator from the structures, particularly the base. Standard mounts that come with generators fail to achieve large attenuations. Springs combined with padding achieve good insulation. In seismic zones, spring supports should avoid displacements.

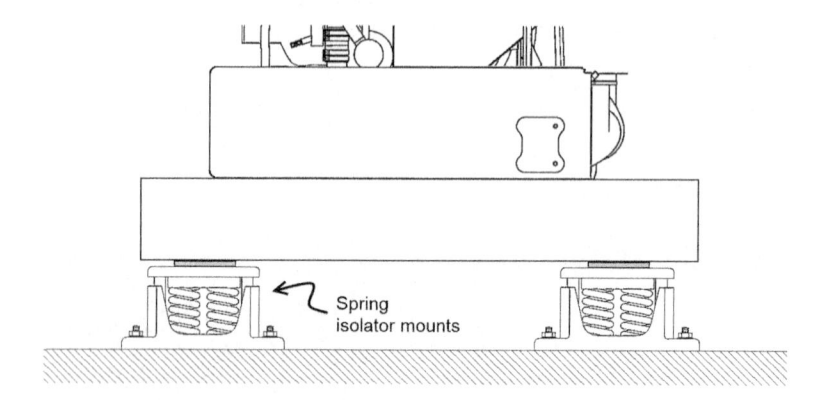

Fig. 4.14. Antivibration spring assembly.

When a large attenuation for highly critical applications is necessary, an inertia base is placed between the floor and the generator and the elements between them are connected with springs. The inertia base is usually a concrete slab of at least 1.5 times the mass of the generator.

It is also necessary to insulate other support points that can transmit vibrations including pipe runs, exhausts, pipe supports and cable supports. When fixing or supporting pipes, arrange the supports and brackets irregularly to help dampen the vibrations:

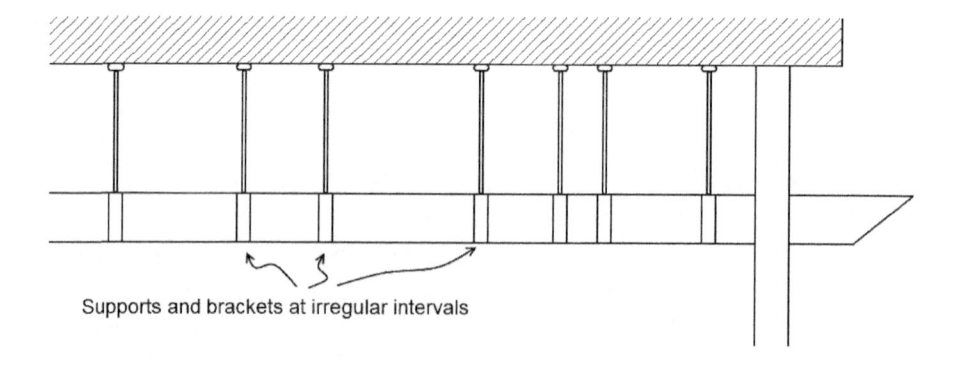

Fig. 4.15. The supports fixed at irregular intervals dampen the vibrations.

Finally, it is important to have a comprehensive approach. There is little point installing very good silencers if you are hardly going to touch the remaining components that will continue to emit a lot of noise!

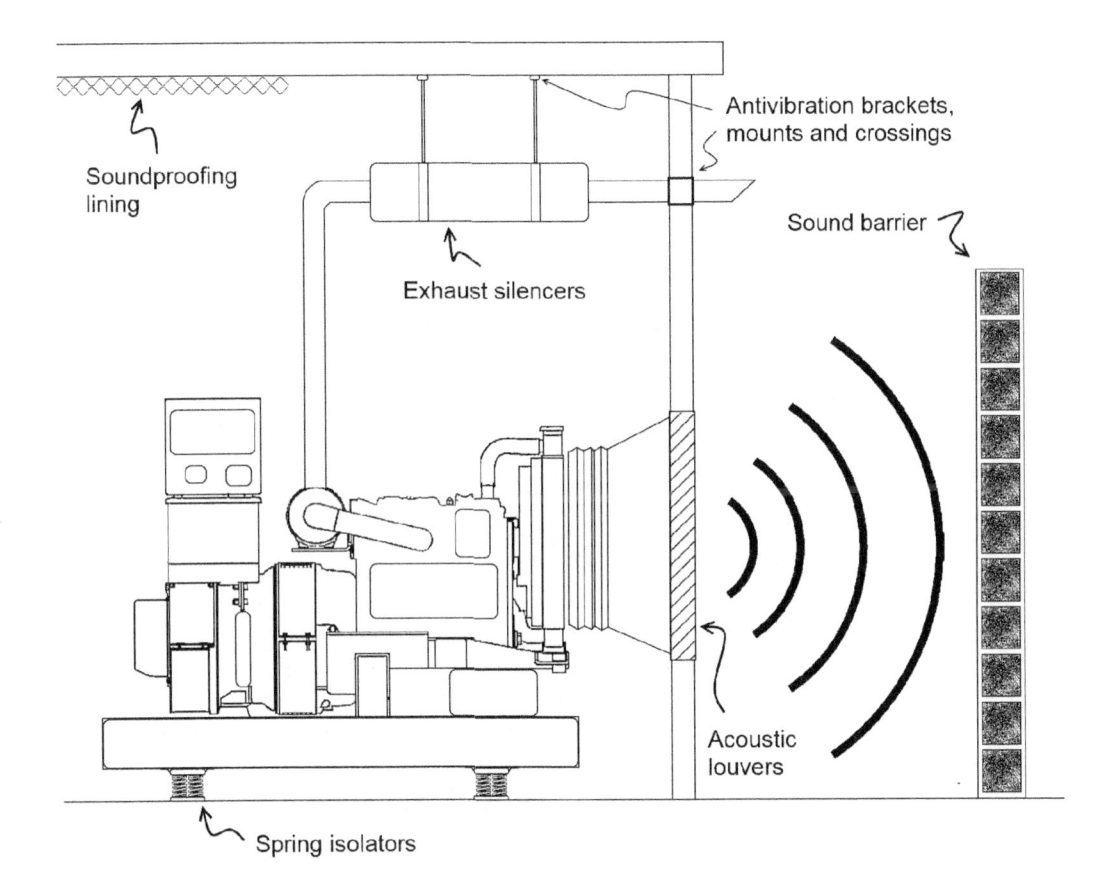

Fig. 4.16. Summary of sound insulation measures.

Electrical protection

⚠ **Electrical work must be impeccable.** Take note on whether the installations are professional or are full of splices, frayed wires and exposed housings as in the picture. Do not allow for installations of this type that endanger people and machines!

Fig. 4.17. Part of the installation of a generator with 380 V (!) cables exposed.

Grounding

⚠ All generators, including portables, must **be grounded** to prevent accidents and to protect appliances.

Grounding is a very simple and inexpensive installation for protection against electrocution. It provides a very low resistance path so that, if leakage occurs, electricity will have an easier passage through the system than through a person. **Fix the grounding before carrying out any operational tests!**

A grounding installation usually consists of one or more metal bars buried and connected to the electrical system through a green and yellow cable. In generators it goes to neutral if there is not a grounding connection, however follow the manufacturer's instructions. All metal housings and metal elements must be connected. Check that there is no other grounding point within 20 metres to avoid interference between them.

Ground fault Interrupter

Installations must have a ground fault interrupter to prevent people being electrocuted and seriously injured or killed.

This protection is assured with a GFI (ground fault interrupter) also known as RCD (residual current device) that breaks the circuit when it detects a leak in it (which is what happens when a person is electrocuted).

The GFI is easily recognized because they usually have a button with a "T" for test. To protect people, the sensitivity must be at least 30 mA.

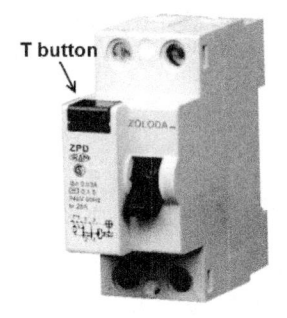

Protection against overloads

If the protection is not included with the generator, it is good practice to fit a magnetothermic breaker at the generator output to prevent overload. A recommended value is commonly 115% protection.

What protection should be fitted at the output of each phase of a 220V/380V generator of 100 kVA? And for a three-phase balanced load and 3 cable conductor?

1. For each phase: $kVA = \dfrac{V*I}{1000}$ Then $I = \dfrac{1.15*kVA*1000}{V}$

$$I = \dfrac{1.15*100*1000}{220} = 522.72 \text{ A.}$$

As only certain nominal currents are manufactured, the next one below is selected at 500 A.

2. For the three-phase balanced system of 3 cables,

$$I = \dfrac{1.15*kVA*1000}{V} = \dfrac{1.15*100*1000}{380} = 302.63 \text{ A} \rightarrow 300 \text{ A.}$$

In annex A are the formulas to calculate powers and currents of different systems. Note that 220 V * 1.732 = 380 V.

Transfer Switches

In its simplest form, a transfer switch allows you to choose the source of electricity that will receive the load and ensures that the two sources are never connected simultaneusly.

Imagine an installation that receives electricity from the mains supply and has a backup generator for when the power goes out:

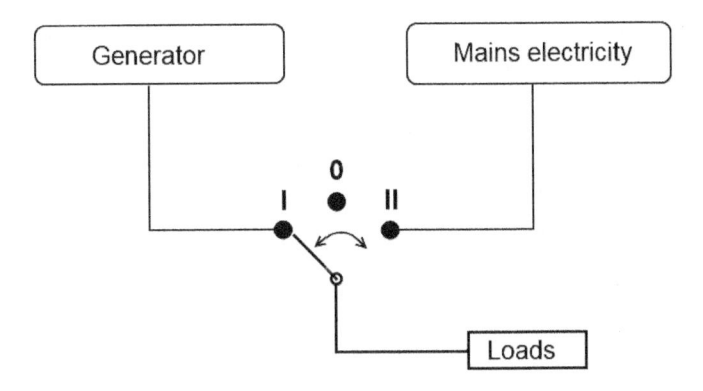

Fig. 4.18. Schematic diagram of a transfer switch setup.

When the switch is in position **I**, the load receives electricity from the generator. When the switch is in position **0**, it is disconnected from the two current sources. And when the switch is in position **II** the load is connected to the mains. When the power cuts out the generator starts, and when the generator is ready to accept the load the switch is turned from position **II** to position **I**.

It is very important to disconnect the transfer switch when electrical work is being carried out, especially with automatic switches, so as to avoid the generator accidentally starting with operators working on it, and to prevent electric shocks.

Unless there is equipment to perfectly synchronize the waves, the generator and any other electricity source must be isolated at all times to avoid catastrophic consequences for all equipment.

Some transfer switches are able to change between sources very quickly. The faster the transition between sources, the more complex the equipment to synchronize the waves. Solid state switches, for example, can make the transfer in less than a ¼ cycle (5 milliseconds). In these cases, it is very important that the waves are synchronized before making the change.

Because transfer switches are fairly specialized equipment and because of the possibility of major damage if they fail, it is probably a good idea to only use to two types of transfers:

- **Automatic transfer switch**, which sends a signal to start the generator and transfers the load.

- **Manual transfer switch,** in which the transfer is made by hand allowing for the generator to start and warm up without load. It usually has a built-in indicator light that is lit when the electricity has come back on:

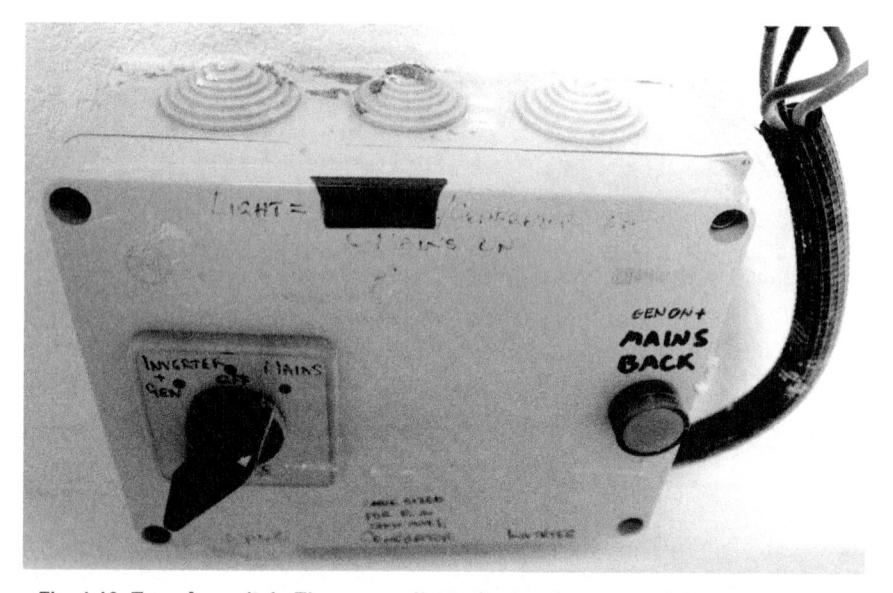

Fig. 4.19. Transfer switch. The green pilot indicates the return of the electricity.

In both cases you can install a small computer UPS for equipment that must not lose power during the change (such as computers, telecommunications equipment or other sensitive equipment). If you still need more sophisticated transfer switches, the manufacturer will be able to help with this.

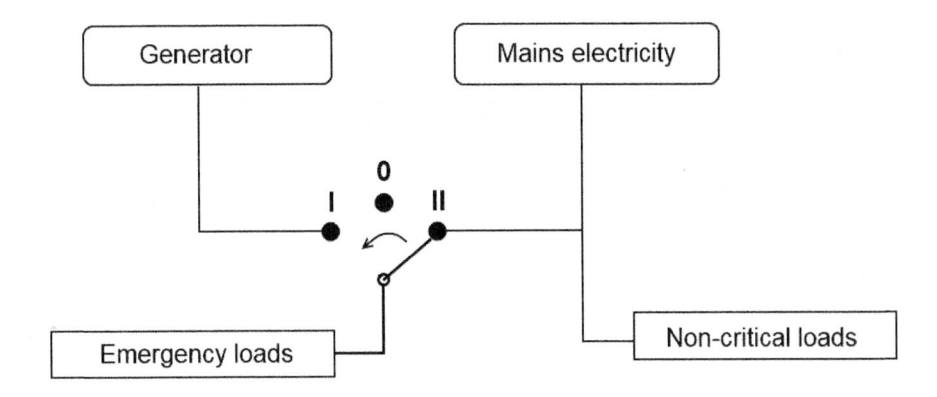

Fig. 4.20. Schematic diagram of a transfer switch setup for emergency loads.

Balancing the loads

In three-phase generators it is very important to distribute the loads evenly between the different phases. To do this, you have to plan the electrical installation so that the circuits of each phase carry as similar a load as possible. When an appliance is three-phase, it consumes in a balanced manner. Not consuming in a balanced way causes overheating and loss of efficiency. In practice, as this is difficult to achieve religiously in small networks, it is important that you bear in mind the following two points:

- **The load of any phase may not exceed the output of the generator.** Just because some phases carry a lighter load does not mean that one can go over the rating.

- **The greater the total load of the generator, the more balanced the phases should be.** The diagram below provides typical tolerance margins of unbalanced loads in generators of less than 200 kW. The phase load should fall within the shaded area, that is, a generator bearing 10% of its power as three-phase load can tolerate an extra load in one of it phases of 60% of the power rating.

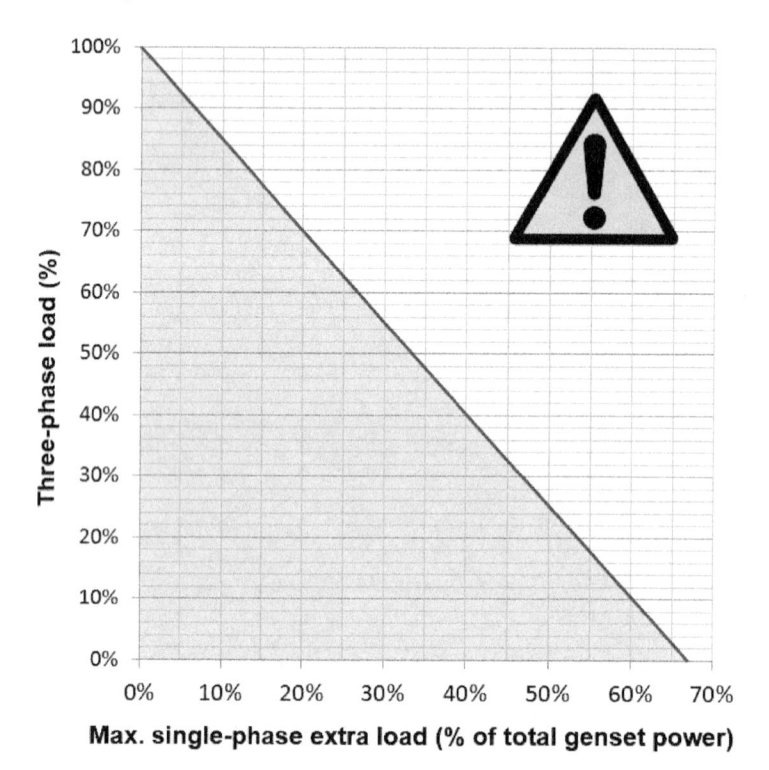

Fig. 4.21. Allowable unbalanced single-phase loads in generators of less than 200 kW.

What is the maximum load of a phase on a 100 kVA generator operating overall at 70%?

1. In the above graph, for a generator with a three-phase overall load of 70% the maximum single-phase extra load is 20%.

2. 20% of the power rating is:

    100 kVA * 0.2 = 20 kVA

    The phase may carry up to 20 kVA in addition to the three phase load.

A generator with unbalanced phase loads **produces imbalanced phase voltages**. Check that these voltages are within the ranges that the equipment can tolerate, especially with three-phase motors.

The voltage regulator of some generators only uses one reference phase to regulate the generator. The loads most sensitive to voltage changes should go in this phase.

Improving existing installations

Now that you know the basic requirements, take a look at the following image for a moment and think about what needs to be improved:

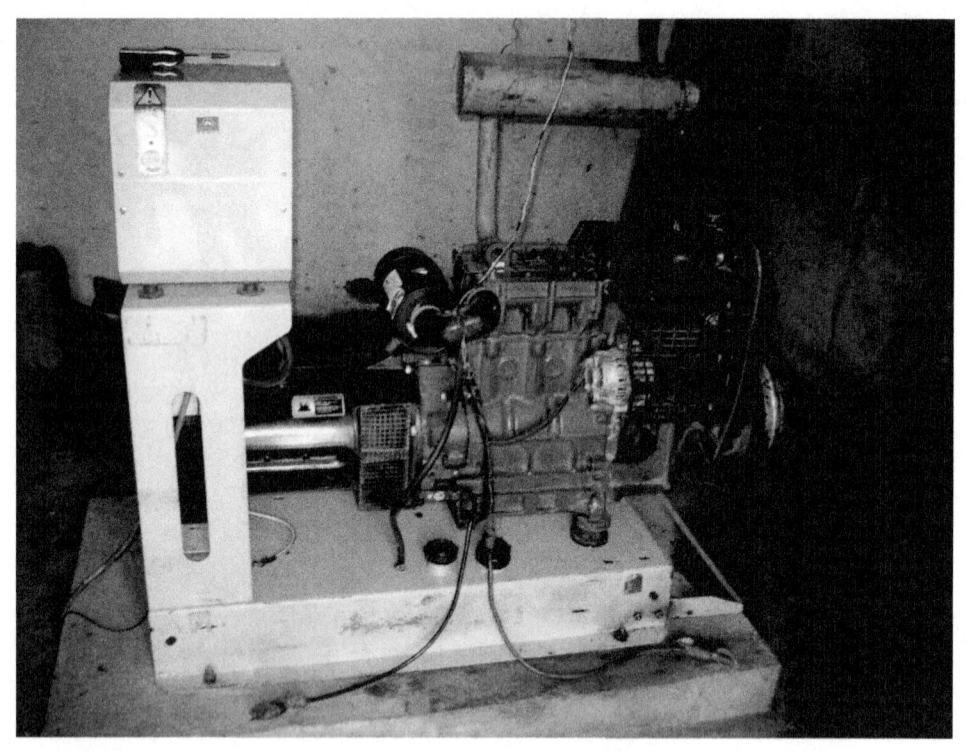

Fig. 4.22. Installation with serious need of improvement.

Here are the three most obvious changes that should be made:

1. *The exhaust is inside the room. Look at the stain on the wall and imagine what the air filter is like. The dirty air filter worsens combustion and makes the generator unstable; in addition, flue gases are highly toxic!!*

2. *The air flows are unorganised. The hot air remains in the room and the generator works overheated.*

3. *Someone took the battery! This often happens in situations where the batteries to jumpstart other machines and various other services. The battery is not only fundamental for the start-up, the generator also needs it to function and some components such as the charger and the voltage regulator can get damaged if it is absent.*

5

Operation and maintenance

Generators should warm up for four to five minutes before the load is applied so that the frequency and voltage stabilizes. These five minutes can be used to record the fuel level, the gauges and to check that everything is working properly.

Similarly, they should also be allowed to cool down without load for another five minutes before turning them off completely. This period can also be used to record operating parameters.

Many other important details will be explained in your generator's instruction manual, which you should read thoroughly before you even think about installing the generator! This chapter focuses on issues that the manual is unlikely to cover.

Record book

Only when you are faced with an unknown generator for which there is no information will you realize how important the records are.

Your record book does not need to be overly detailed in order to be useful. In most cases, a simple logbook that records incidents, maintenance and observations according to the generator's operational hours will be a big help. For example:

Date	Hours	Note	Cost
19-11-08	50	End of running-in, without incident. Change oil and filter	$100
20-02-09	250	250 hr maint. as per maintenance manual	$164
25-03-09	367	Clogged fuel filter. Fuel purchased at RBS	
26-06-09	472	Damaged fan belt, tensioner seized up.	$168
……..	…..	………….	……..
17-12-11	4367	Noise in alternator bearing. Re-greased and monitored	

Fig. 5.1. Example of a simple record book.

In addition to this essential record book you can keep additional records depending on the situation. Two records books that may be useful are:

- A use monitoring book with start times, stop times, diesel refueling, voltage, frequency, etc. An added advantage of these records is that, given they are maintained on a daily basis, it promotes the care of the generator and gives a sense of supervision.

- A collection of signed sheets with a list of periodic maintenance tasks corresponding to the hours of service where the operations carried out are marked. The idea is that this clearly establishes responsibility, and avoids the lack of control regarding what should have been done and what was actually done.

Measuring diesel consumption

Measuring diesel consumption is essential to evaluate the performance of the engine and to also check for the often-present issue of possible fuel theft. To measure the use of fuel in generators, follow this process:

1. Select a load that will be consistent and that represents a significant percentage of the power of the generator, at least 30% and preferably 50-60% of what is typically used by manufacturers. A three-phase load is best if the generator is three-phase, to avoid the complications of balancing the load. Isolate the rest of the loads so they do not become operational accidentally or unknowingly.

2. Find a safe and suitable sized container for the duration of the proposed test. Fill it with fuel and change the feed pipe so that the generator is supplied from this container.

3. Weigh the container with the fuel. The weight gives a more accurate measurement than a volume estimate. Weigh a litre of fuel to get the density since it is variable around 850 g/litre according to the type of fuel and the region.

4. Record the time, start the generator up and connect the load.

5. Record:

    - The temperature throughout the test at regular intervals.
    - The electricity generated at regular intervals, either with a kWh meter or recording current and voltage to calculate the power.
    - The weight of the fuel tank if this is possible and practical.

6. Run the generator with the load for a significant time period, at least several hours. The more time you record within that which the situation allows, the more accurate the results.

7. Turn the generator off, record the time and weigh the container with the fuel. The weight of the container itself does not influence the calculations.

8. Calculate the results and compare them with the data from the first chapter.

9. Be sensible about the findings. Do not rush to accuse someone of stealing or buy a new generator in the event that you find a problem.

Try to ensure that the measuring devices that you use are accurate to at least ± 2%.

If you want to settle accusations of theft, then you will need an outside person to oversee the process. Be aware that there are many factors affecting consumption that have nothing to do with greed. Just because a generator can work with the expected consumption, does not mean that it will work with the same consumption when used unwisely, as for example, without controlling the appliances connected or with very low loads. On many occasions you can avoid running the risk of unjustly accusing someone by instead establishing supervision and doing away with the lack of control. Once this happens, the fuel stealing usually ends.

A 25 kW generator has been tested for 6 hours to verify its consumption with a three-phase submersible pump with the following results:

> **Density of the diesel used: 850 g/l**
> **Starting weight: 40 kg**
> **Final weight: 13 kg**
> **Voltage: 380 V**
> **Current: 19.2 A**

1. The three-phase power is calculated with the following formula:

    $$P = I * \sqrt{3} * V * \cos \Phi$$

    For a power factor (cos Φ) of 0.8, the formula is:

    $$P = I * V * 1.384 = 19.2 * 380 * 1.384 = 10097.66 \text{ W} \approx 10.1 \text{ kW}$$

2. This power consumed for 6 hours corresponds to an energy of:

    10.1 kW * 6 hours = 60.6 kWh

3. The generator load has been: $\dfrac{10.1 \text{ kW}}{25 \text{ kW}} = 40.4\%$

CHAPTER 5. Maintenance

4. The diesel consumption has been: $\dfrac{40 \text{ kg} - 13 \text{ kg}}{0.85 \text{ kg/l}} = 31.76$ litres

5. The consumption per kWh has been: $\dfrac{31.76 \text{ litres}}{60.6 \text{ kWh}} = 0.524$ l/kWh

Comparing the consumption with that expected, approximating to the graph from the first chapter, it seems high.

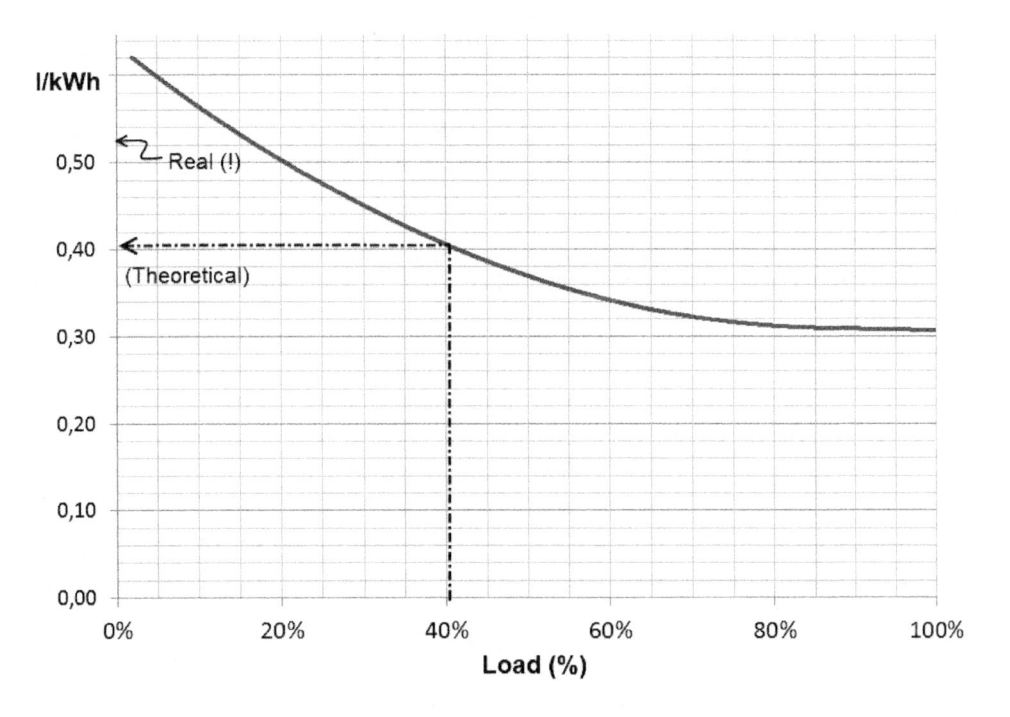

Evaluating the quality of diesel

The majority of manufacturers include diesel specifications in their user manuals, however analysing these features is often impossible and expensive. The fuel passes through many people and it is often small retailers in a garage who end up selling it. Even if it were possible to analyse, it is unlikely that the batches of fuel, even from the same source, would have the same characteristics and origin from one to another.

Even so, you can evaluate the quality of the diesel to some extent by doing some simple tests:

- Keep track of consumption in relation to the load.

- Fill a transparent container with fuel and let it settle. Fuels which are contaminated with water, sand or sediment will form a layer on the bottom and those fuels contaminated with gasoline or kerosene, will form a layeron the surface.

Try to ensure that your diesel supply is trustworthy and stable. If your generator's fuel tank is corroded, the pump and injectors are giving you trouble, the filters are clogged or sludge has formed, you have been using poor quality fuel.

The use of **biodiesel** or a mixture thereof is generally not recommended due to the decrease in power and the increase in consumption.

Overhaul

An overhaul involves dismantling the engine to inspect it, replacing some parts and rebuilding others to return the generator to its original performance level.

In addition to accounting for up to 30% of the cost of the generator over its lifespan, undertaking an overhaul is a laborious process that takes time and requires organization. One decision that must be made in advance is how or whether to organize an alternative source of electricity while the overhaul is carried out.

Make sure that the workshop responsible for the service has all the necessary tools and replacement parts actually available. You do not want to wait six months for something to arrive from Dubai, while all the pieces are on the floor!

Deciding when to do it

The manufacturer of the generator provides generic limits but it really depends on each machine and its use. The need for an overhaul can be determined upon:

- The number of hours and the operating load.
- The total amount of fuel consumed.
- An increase in the noise and vibration.
- An increase in the oil consumption.
- Analysis of the metal fragments in the oil.

Postponing the overhaul in order to save money is a bad idea. This is because:

- It increases the chances of a catastrophic failure that will finish off the generator or that will require a large number of parts and hours of work.

- When the wear is excessive, far fewer parts can be reused.

If **the lubricating oil consumption has tripled** due to normal engine wear only, organize the overhaul right away. If maintenance records are kept, it is easy to determine what consumption is normal, and whether you have deviated from this. If these records don't exist, you may assume an oil consumption of between 0.1 to 1.6 g/kWh produced, or about 0.1% of the fuel consumed (as an indication, because it varies depending on the age, power and load of the engine).

In development contexts it is common to come across a generator that does not have much history, which leaves you with very little information to decide what is to be done with it. Investigate oil consumption. Sometimes the performance alone gives you an orientation. Another indication is a bluish-grey smoke that indicates the engine is burning oil.

Fig. 5.2 This generator needs a service and an hot air discharge duct!

Types of overhaul

There are two types of overhaul:

- **Top end overhaul,** in which only the cylinder head is dismantled and inspected, rebuilding or replacing the parts that are visible.

- **Complete overhaul,** in which the motor is completely dismantled.

The typical tasks for the two types of overhauls are listed at Annex F.

Remember that after each overhaul the generator must return to its cycle, changing the oil at 50 hours.

Long term storage

If the generator is not going to be used for over 30 days, or cannot be installed immediately, follow the following process as well as the manufacturer's recommendations. As always, this should only be undertaken by qualified personnel because it is potentially dangerous.

Before storage

- Completely empty:
    - the fuel tank
    - the engine oil
    - the coolant.
- Disconnect it from the network.
- Disconnect the battery cables.
- Clean and protect the generator from dust with a tarpaulin if it is an open generator.
- In humid places, place desiccant bags in the radiator grille and in the exciter.

During storage

The storage location should be a clean place, with no extreme temperatures, vibration or other machines running that damage the bearings, no dust or humidity and no bugs or rodents. It is desirable to:

- Manually turn the rotor of the alternator every month or anticipate a change of bearings.
- Charge the battery periodically or plan its replacement.

Removing from storage

- Check the insulation of the coils with a megohmmeter.
- Top up the fuel, coolant and oil.
- Change the oil filter if it hasn't been done or it has been in storage for a long time.
- Connect the batteries after fully charging.
- Warm up the motor and load lightly, checking there are no leaks.

As some of these tasks may prove to be quite impractical, perhaps you should ask yourself if you can instead simply leave it where it is and run it for a short period once a month.

With small portable gasoline generators, such below, the gasoline which evaporates from the carburetor after extended periods of inactivity prevents them from starting later. To avoid getting thrown off at the time when they are most needed, the stop valve for the gasoline can be shut with the engine running, or the generator can be kept running until it runs out of gasoline to make sure the carburetor is completely empty before storage.

Fig. 5.3. Small gasoline portable generator (Pramac E3200).

Operation in adverse conditions

Very cold environments

A generator, the installation and its operation need to be adapted for very cold climates. Try to learn from local experience what typcial problems might be, and how to fix them. The following are some of the problems you can expect and their solutions.

Installation precautions:

- If the generator arrives during very cold weather, let it adjust to the room temperature for 24 hours before removing the packaging and protection to prevent condensation.

- Generators move a significant volume of air during operation and so if it is very cold, installations in the same building can quickly freeze. Be careful of things that may be sensitive to freezing.

- Diesel gellifies with the cold. The installation will probably require some method of heating the fuel lines.

- Consider the effects of cold on the cables, protections and other sensitive parts.

- Prevent ice and snow from clogging the air vents by using protectors or closing louvers.

- To prevent the air filter from freezing, the inlet air can be mixed with warm air from a hot air duct through a regulated recirculation loop.

- The magnetothermic circuit breakers need more current to activate when it's cold. So a switch that cuts out at 15 A for 40 °C, will cut out at 17 A for 25 °C. With very low temperatures the threshold increase can be very important.

- Exhaust pipes which are fitted with a rain cap may still freeze in a closed position.

Precautions during operation:

- Diesel becomes cloudy with the cold and can clog the fuel filter. In this case use diesel suitable for the cold; diesel type No. 1-D does not cloud until -26 ºC while the No. 2-D does at -6 º C.

- Avoid turning the generator on and off too frequently to prevent wear and damage to the valves due to carbon deposition.

- Add antifreeze in the proportions recommended by the manufacturer without exceeding 50% to prevent the engine from overheating. Even in very hot climates at least 5% of antifreeze is necessary to prevent corrosion of the engine.

- You may need to use water heaters, battery heaters, and engine block heaters.

- Fill the fuel tank to the top after each use to prevent condensation.

- The batteries have less power and engines offer much more resistance when starting up. At -20°C, a battery has half the power than at 25°C and the engine needs eight times more power to start up. Batteries must be well charged and possibly oversized. Load measurements using a hydrometer need corrections at low temperatures.

- The hot air discharge duct can have a hatch in the top to discharge the hot air into the room on start-up or at low loads that produce an excesive cooling.

Coastal environments

Coastal environments are considered to not only be directly on the coast, but also those places within 100 km of a body of saline water - and 100 km is a long way! The two main problems associated with coastal environments are moisture and corrosion.

- Choose generators for marine duty when possible.

- In humid environments, it is recommended to install heaters in the alternator to maintain the temperature at 5 °C above ambient.

- Convoluted inlets in humid areas promote the precipitation of moisture.

- Closing the inlets and outlets after switching off the generator will reduce condensation.

- The cabins and parts of the enclosure should be galvanized, aluminum or painted with salt-resistant paint.

Dry and arid environments

In dry and arid environments, problems are often caused by the dust and sand. Some precautions that may be helpful are:

- Install the generator so that it is protected from blowing sand. The sand carried by the wind can wear down the paint and insulation causing serious damage.

- The cooling fins and the radiator should not be close together, to avoid trapping dirt and sand.

- Oversizing of the cooling capacity to 115% is recommended.

- Prevent contamination of the diesel by sand or dust.

- A higher class of insulation can prevent premature deterioration of the insulation from overheating of parts with dirt and dust that do not dissipate heat as effectively.

- In very hot climates, install exhausts with thermal insulation. As an example, each metre of 3" exhaust pipe emits 5 kW of heat. An exhaust of only two metres is equivalent to having five 2 kW heaters running at full power!

At high altitude

- In addition to derating of the generator, some components such as circuit breakers will also require derating.

- The radiator becomes less efficient due to the lower density of the air. Check that the radiator from the factory is suitable for the actual altitude where the generator will be installed.

- The air flow rate needs to be increased by 9% per 1000 m. In some extreme cases, you may have to install additional fans.

arnalich

water and habitat

To engage **consulting services** for generators and renewable energy, write to contact@arnalich.com.

Annexes

A. CALCULATION OF AC POWER

The power measured in kVA, where V is the voltage, and I is the current, is:

Single-phase system:

$$kVA = \frac{V * I}{1000}$$

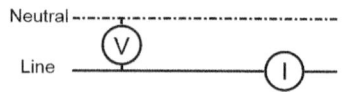

Balanced 3 cable three-phase system:

$$kVA = \frac{V * I * 1.732}{1000}$$

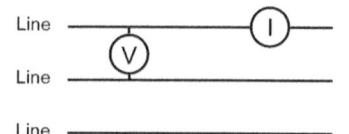

Unbalanced 3 cable three-phase system:

$$kVA = \frac{V * \frac{I_1 + I_2 + I_3}{3} * 1.732}{1000}$$

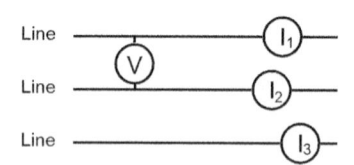

Balanced 4 cable three-phase system:

$$kVA = \frac{V * I * 3}{1000}$$

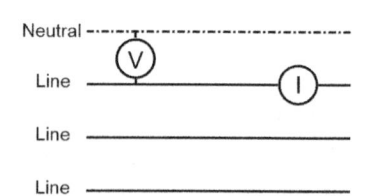

Unbalanced 4 cable three-phase system:

$$kVA = \frac{V * \frac{I_1 + I_2 + I_3}{3} * 3}{1000}$$

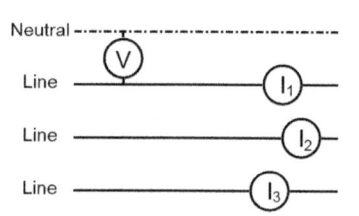

© Santiago Arnalich   Arnalich. Water and habitat   www.arnalich.com

B. USEFUL LIFE OF A GENERATOR

Generators in acceptable working conditions and with proper maintenance generally last between 20,000 to 30,000 hours with an overhaul between 10,000 and 15,000 hours. So don't be surprised if a generator that is used 22 hours a day has to be replaced after only two and a half years!

Hours of use per day	Lifespan
1	54.8
2	27.4
3	18.3
4	13.7
5	11.0
6	9.1
7	7.8
8	6.8
9	6.1
10	5.5
11	5.0
12	4.6
13	4.2
14	3.9
15	3.7
16	3.4
17	3.2
18	3.0
19	2.9
20	2.7
21	2.6
22	2.5

C. GRUNDFOS SUBMERSIBLE PUMPS START-UP PEAK

Electric motors have a peak consumption during start up. The pump manufacturer will tell you what this peak consumption is. When there is a lack of data use a factor of 3. As a guide, these are the factors recommended by Grundfos, Spain.

Motor power of the pump		Minimum power of the generator		Factor
HP	kW	kW	kVA	
4	3	8	10	2.7
5.5	4	10	12.5	2.5
7.5	5.5	12.5	15.6	2.3
10	7.5	15	18.8	2.0
12.5	9.2	18.8	23.5	2.0
15	11	22.5	28	2.0
17.5	12.8	26.4	33	2.1
20	15	30	37.5	2.0
25	18.5	40	50	2.2
30	22	45	56.5	2.0
35	26	52.5	65	2.0
40	29.5	60	75	2.0
50	37	75	94	2.0
60	44	90	112.5	2.0
70	51.5	105	131	2.0
80	59	120	150	2.0
90	66	135	170	2.0
100	73.5	150	190	2.0
125	92	185	230	2.0
150	110	210	260	1.9

D. APPROXIMATE WEIGHTS AND DIMENSIONS OF GENERATORS

Prime power (kVA)	Weight (Kg)	Length (m)	Width (m)	Height (m)
8.5	334	1.3	0.6	1.2
12.5	393	1.3	0.6	1.2
16.5	454	1.3	0.6	1.2
20	467	1.3	0.6	1.2
27	800	1.8	0.7	1.4
30	810	1.8	0.7	1.4
40	890	2.1	0.8	1.4
45	910	2.1	0.8	1.4
50	910	2.1	0.8	1.4
60	960	2.1	0.8	1.4
80	1010	2.1	0.8	1.4
100	1180	2.4	0.7	1.4
135	1417	2.7	0.9	1.5
150	1535	2.7	0.9	1.6
180	1663	2.8	0.9	1.6
200	2052	3.0	1.0	1.7

E. APPROX. DIMENSIONS OF INSTALLATION ELEMENTS

These dimensions are intended as guidance only. Always use the manufacturer's recommendations whenever possible.

Stand/by power (KVA)	Generator Length (m)	Generator Width (m)	Generator Height (m)	Generator room Length (m)	Generator room Width (m)	Generator room Height (m)	Hot air outlet Length (m)	Hot air outlet Width (m)	Hot air outlet Area (m^2)	Air inlet Area (m^2)	Door Width (m)	Door Height (m)	Exhaust Diameter "
11	1.6	0.8	1.1	3.5	3.0	2.7	0.7	0.8	0.5	0.8	1.5	2.2	3.0
14	1.2	0.6	1.0	3.5	3.0	2.7	0.7	0.8	0.5	0.8	1.5	2.2	3.0
16	1.1	0.6	0.9	3.5	3.0	2.7	0.7	0.8	0.5	0.8	1.5	2.2	3.0
20	1.9	0.9	1.2	3.5	3.0	2.7	0.7	0.8	0.5	0.8	1.5	2.2	3.0
25	1.9	0.9	1.2	3.5	3.0	2.7	0.7	0.8	0.5	0.8	1.5	2.2	3.0
27	1.2	0.6	0.9	3.5	3.0	2.7	0.7	0.8	0.5	0.8	1.5	2.2	3.0
30	1.9	0.9	1.3	3.5	3.0	2.7	0.8	0.8	0.5	0.8	1.5	2.2	3.0
33	1.9	0.9	1.2	3.5	3.0	2.7	0.7	0.8	0.5	0.8	1.5	2.2	3.0
40	1.7	0.9	1.4	3.5	3.0	2.7	0.7	0.8	0.5	0.8	1.5	2.2	3.0
43	1.9	0.9	1.4	3.5	3.0	2.7	0.8	0.8	0.5	0.8	1.5	2.2	3.0
47	1.7	0.9	1.2	3.5	3.0	2.7	0.8	0.8	0.5	0.8	1.5	2.2	3.0
50	1.7	0.9	1.4	3.5	3.0	2.7	0.8	0.8	0.5	0.8	1.5	2.2	3.0
66	1.9	0.9	1.5	3.5	3.0	2.7	0.8	0.8	0.5	0.8	1.5	2.2	3.0
70	1.9	0.9	1.8	3.5	3.0	2.7	0.8	0.8	0.5	1.0	1.5	2.2	3.0
80	1.7	0.9	1.3	3.5	3.0	2.7	0.8	0.9	0.5	1.0	1.5	2.2	3.0
88	1.9	0.9	1.3	3.5	3.0	2.7	0.8	0.9	0.5	1.0	1.5	2.2	3.0
93	2.2	1.0	1.6	4.0	3.0	2.7	0.8	0.9	0.5	1.0	1.5	2.2	3.0
101	2.2	1.0	1.4	4.0	3.0	2.7	0.8	0.9	0.5	1.0	1.5	2.2	3.0
110	2.2	1.0	1.5	4.0	3.0	2.7	0.8	0.9	0.5	1.0	1.5	2.2	3.0
125	2.2	1.1	1.7	4.0	3.0	2.7	1.1	1.0	0.6	1.0	1.5	2.2	3.0
145	2.3	1.0	1.5	4.0	3.0	2.7	0.8	0.9	0.5	1.0	1.5	2.2	3.0
150	2.2	1.1	1.7	4.0	3.0	2.7	1.1	1.0	0.6	1.0	1.5	2.2	3.0
154	2.3	1.0	1.5	4.0	3.0	2.7	1.1	1.0	0.6	1.3	1.5	2.2	3.0
165	2.3	1.0	1.5	4.0	3.0	2.7	1.1	1.0	0.6	1.3	1.5	2.2	3.5
175	2.3	1.1	1.68	4	3	2.7	1.1	1	0.6	1.3	1.5	2.2	4.5
200	2.3	1.1	1.68	4	3	2.7	1.1	1	0.6	1.3	1.5	2.2	4.5

F. TYPICAL TASKS INVOLVED IN AN OVERHAUL

This is not an exhaustive list, and not all tasks may be applicable depending on the generator. The idea of these lists is to guide you when you are faced with a service negotiation or when drafting an overhaul contract.

Top end overhaul:

1. Inspection and machining of the valve and seat.
2. Inspection and possible replacement of the valve guides.
3. Dismantling and cleaning of the turbo.
4. Replacement of the turbo bearings if necessary.
5. Inspection, maintenance and / or calibration of the injection pump(s).
6. Cleaning carbon from cylinders and pistons.
7. Inspection of the sump to detect traces indicating unusual wear.
8. Replacing the injectors, depending on the use

Complete Overhaul:

In addition to the tasks listed above for a top end overhaul, a complete overhaul also requires the following additional activities:

1. General inspection of items for cracking, deformation, corrosion and deposits: camshaft, camshaft shock absorbers, cylinder block, cylinder, pistons and connecting rods, cylinder head, flywheel, and other associated parts.
2. Inspection and / or reconstruction of the rockers of the exhaust valves.
3. Inspection, reconstruction or replacement of connecting rods, cylinder head assembly, injection pumps, oil cooler, piston pins, and other associated parts.
4. Inspection and / or replacement of: engine mounts, cylinder liners, piston skirts and heads, engine wiring, and other associated parts..
5. Replace bearings and camshaft seals, injectors, main bearings, exhaust and air intake gaskets, and other associated parts.

Bibliography

- *Application engineering. T-030: Liquid-Cooled Generator Set Application Manual.* Cummings Power Generation 2012.
- Bolton, Paul (2013). *Petrol and diesel prices.* Standard Note SN/SG/4712. Library of the House of Commons.
- *Diesel Generator Operation and Maintenance Manual.* Hyundai 2011.
- *Diesel Generator Group Operating and Maintenance Manual.* EMSA 2008.
- Diesel Generating Sets Installation recommendations and Operations Manual. AKSA 2012.
- *Diesel Generator Set Model DGBC 60 Hz Specification sheet.* Cummings 2006.
- *Cold Climate Considerations for Generator Set.* Infosheet #32. Clifford Power Systems 2008.
- *Cold Weather Recommendations for all Caterpillar Machines.* Caterpillar 2007.
- *Mahon, L.L.J (1992). Diesel generator handbook.* Newnes.
- *Electric Power Applications, Engine & Generator Sizing. Application and Installation GuideCaterpillar.* Caterpillar 2008.
- *Fossil Fuel Price Projections.* DECC, UK Government 2012.
- *Generator Noise Control - An Overview.* Ashrae Technical Committee on Noise and Vibration 2002.
- *Generator Set Installation Guidelines.* Generator Joe 2013.
- *Generator Set Installation Recommendations.* Baldor. 2005.
- *Generator set noise solutions: Controlling unwanted noise from on-site power systems.* PT 7015. Cummings 2007.
- *Generator Set Ratings Guidelines.* TIB 101. Kohler Power Systems 2001.
- *Generator Sizing Guide.* TD00405018E. Eaton Power.
- Generator User Manual. Models S3100 – S5000 – S7500 – S10000 –S12000. Pramac Power Systems.
- *Grounding of AC generators and switching the neutral in emergency and standby power systems.* PT 6006. Cummings 2006.
- *How to size a genset: Proper generator set sizing requires analysis of parameters and loads.* PT 7007. Cummings 2007.
- *Guide to Generator Set Exhaust Systems.* Infosheet #16. Clifford Power Systems 2008.
- *Installation and Maintenance Manual Synchronous generators.* Weg Industrias, 2003.

- *IP code*. Wikipedia.
- *ISO 8528-1. Reciprocating internal combustion engine driven alternating current generating sets. Part 1: Application, ratings and performance.* ISO 2005.
- Marfell, Ray. (2007). *Demistifying generator set ratings.* Caterpillar.
- Mestre, J. (2008.) *NTP 142: Grupos electrógenos. Protección contra contactos eléctricos indirectos.* INSHIT.
- *Operation and controller manual.* Bindu Power 2012.
- *Operation and Maintenance Manual: 3500 Generator Sets.* Caterpillar 2009.
- *Operation and Maintenance Manual: SRB4 Generator Sets.* Caterpillar 2000.
- *Operation, Maintenance and Repair of Auxiliary Generators.* TM 5-685/NAVFAC MO-912. US Army. 1996.
- *P7 AC Generators. Installation, Servicing and Maintenance.* Stamford 2011.
- *Sound advice on attenuating genset noise, vibrations.* Webarticle by Ken Lovorn. CSEMAG.
- *The True Cost of Providing Energy to Telecom Towers in India.* Intelligent energy 2012.
- *Transfer switch application manual.* Cummings 2004.
- *Understanding load factor implications for specifying onsite generators.* MTU Onsite Energy 2011.
- *Use and Maintenance manual. GSW Generators.* Pramac Power Systems 2013.
- *World Energy Assessment: Energy and the Challenge of Sustainability.* UNDP 2000.

Pramac branches worldwide

Pramac Energy Generation has sponsored the translation of this manual into english.

Ω PRAMAC

- Production and Commercial Branch
- Commercial Branch

Worldwide

From Italy, worldwide. Our service through a global network to be closer to you.
For further information: www.pramac.com

EUROPE

Italy
PR INDUSTRIAL s.r.l.
Headquarters:
Località Il Piano
53031 Casole d'Elsa, Siena
Tel.: +39 0577 9651
Fax: +39 0577 949076

Germany
PRAMAC GmbH
Salierstr. 48
70736 Fellbach, Stuttgart
Tel.: +49 711 517 4290
Fax: +49 711 517 42999

Spain
PRAMAC IBÉRICA, S.A.
Parque Empresarial Polaris
C/ Mario Campinoti, 1
Av. Murcia-San Javier, km. 18
30591 Balsicas, Murcia
Tel.: +34 968 334 900
Fax: +34 968 579 321

United Kingdom
PRAMAC UK, Ltd.
Crown Business Park, Dukestown
Tredegar, NP22 4EF
Tel.: +44 1495 713 300
Fax: +44 1495 718 766

France
PRAMAC FRANCE S.A.S.
Place Léonard de Vinci
42190 St. Nizier sous Charlieu
Tel.: +33 (0) 477 692 020
Fax: +33 (0) 477 601 778

Poland
PRAMAC Sp z.o.o.
ul. Krakowska 141-155 budynek F
50-428 Wrocław
Tel.: +48 71 7822690
Fax: +48 71 7981006

Romania
S.C. PRAMAC Group S.R.L.
Sos Bucaresti
Targoviste Nr 12A, Corp. A, Etaj 3
077135 Mogosoaia, Ilfov
Tel.: +40 31 417 07 65
Fax: +40 31 417 07 55

Russian Federation
PRAMAC-RUS Ltd
Neverovskogo street 9,
office 316
Moscow
Tel.: +7 985 651 68 66
Fax: +7 985 651 68 66

NORTH AMERICA

PRAMAC AMERICA, LCC
North America
1300 Gresham Road - Marietta, GA 30062
Tel.: +1 770 218 5430
Fax: +1 770 218 2810
Toll Free: +1 888 977 2622 (9 PRAMAC)

PRAMAC INDUSTRIES, INC
C. America, Caribbean
& Andean countries
1300 Gresham Road - Marietta, GA 30062
Tel.: +1 770 218 5430
Fax: +1 770 218 2810

SOUTH AMERICA
& CARIBBEAN

Dominican Republic
PRAMAC CARIBE C. por A.
Avda. 27 de Febrero, Esq. Caonabo,
664 Los Restauradores
10137 Santo Domingo
Tel.: +1 809 531 0067
Fax: +1 809 531 0273

Brazil
PRAMAC BRASIL
EQUIPAMENTOS LTDA.
Av. Victor Andrews, 3210
Bairro Éden - Cep 18086-390
Sorocaba, São Paulo
Tel.: +55 15 3412 0404
Fax: +55 15 3412 0400

ASIA

United Arab Emirates
PRAMAC MIDDLE EAST FZE
1206 JAFZA View 18, P.O.Box 262478
Jebel Ali Free Zone - South 1, Dubai
Tel.: +971 4 8865275
Fax: +971 4 8865276

Singapore
PRAMAC (ASIA) PTE LTD.
2, Tuas View Place
#01-01 Enterprise Logistics Center
637431 Singapore
Tel.: +65 6558 7888
Fax: +65 6558 7878

AFRICA

Senegal
PRAMAC LIFTER
AFRIQUE TRADING S.a.r.l.
Route de l'Aéroport x VDN
B.P. 8959 Dakar
Tel.: +221 33 869 3121
Fax: +221 33 820 8598

Commercial partners
www.pramac.com

Version 1.0

Printed in Great Britain
by Amazon